P9-CEY-761

ALSO BY MARTIN GARDNER

MATHEMATICAL GAMES

The Scientific American Book of Mathematical Puzzles and Diversions

The Second Scientific American Book of Mathematical Puzzles and Diversions

New Mathematical Diversions

The Numerology of Dr. Matrix

The Unexpected Hanging

The Sixth Book of Mathematical Games from Scientific American

SCIENCE, PHILOSOPHY, AND MATHEMATICS

Fads and Fallacies in the Name of Science

Logic Machines and Diagrams

Relativity for the Million

The Ambidextrous Universe

Mathematics, Magic, and Mystery

Great Essays in Science (editor)

LITERARY CRITICISM

The Annotated Alice

The Annotated Snark

The Annotated Ancient Mariner

The Annotated Casey at the Bat

FICTION

The Flight of Peter Fromm

Mathematical Carnival

MARTIN GARDNER

Mathematical

ALFRED A. KNOPF
New York 1975

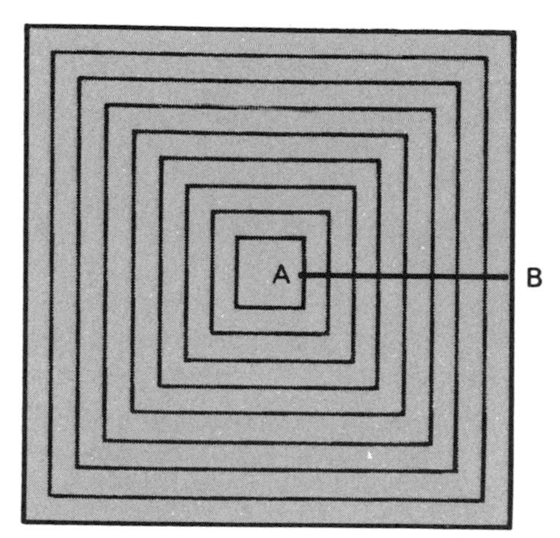

Carnival

From Penny Puzzles, Card Shuffles,

and Tricks of Lightning Calculators

to Roller Coaster Rides into the Fourth Dimension.

A new round-up of Tantalizers

from *Scientific American*

with Elegant Mathematical Commentaries

(and Afterthoughts) by Mr. Gardner,

Backtalk from Readers,

and 115 Pictures and Diagrams.

THIS IS A BORZOI BOOK
PUBLISHED BY ALFRED A. KNOPF, INC.

 Published in the United States by Alfred A. Knopf, Inc., New York, and simultaneously in Canada by Random House of Canada Limited, Toronto. Distributed by Random House, Inc., New York.

Most of this book originally appeared in slightly different form in Scientific American.

Library of Congress Cataloging in Publication Data

*Gardner, Martin, [Date]
Mathematical carnival.
Bibliography: p.
1. Mathematical recreations. I. Title.
QA95.G286 793.7'4 75-8208
ISBN 0-394-49406-7*

Manufactured in the United States of America

PUBLISHED OCTOBER 9, 1975
SECOND PRINTING, OCTOBER 1975

To
John Horton Conway,
whose continuing contributions
to recreational mathematics
are unique in their combination
of depth, elegance, and humor

Contents

Introduction

A TEACHER of mathematics, no matter how much he loves his subject and how strong his desire to communicate, is perpetually faced with one overwhelming difficulty: How can he keep his students awake?

The writer of a book on mathematics for laymen, no matter how hard he tries to avoid technical jargon and to relate his subject-matter to reader interests, faces a similar problem: How can he keep his readers turning the pages?

The "new math" proved to be of no help. The idea was to minimize rote learning and stress "why" arithmetic procedures work. Unfortunately, students found the commutative, distributive, and associative laws, and the language of elementary set theory to be even duller than the multiplication table. Mediocre teachers who struggled with the new math became even more mediocre, and poor students learned almost nothing except a terminology that nobody used except the educators who had invented it. A few books were written to explain the new math to adults, but they were duller than books about the old math. Eventually, even the teachers got tired of reminding a child that he was writing a numeral instead of a number. Morris Kline's book, *Why Johnny Can't Add*, administered the *coup de grâce*.

The best way, it has always seemed to me, to make mathematics interesting to students and laymen is to approach it in a spirit of play. On upper levels, especially when mathematics is applied to practical problems, it can and should be deadly serious. But on lower levels, no student is motivated to learn advanced group theory, for example, by telling him that he will find it beautiful and stimulating, or even useful, if he becomes a particle physicist. Surely the best way to wake up a student is to present him with an intriguing mathematical game, puzzle, magic trick, joke, paradox, model, limerick, or any of a score of other things that dull teachers tend to avoid because they seem frivolous.

No one is suggesting that a teacher should do nothing but throw entertainments at students. And a book for laymen that offers nothing but puzzles is equally ineffective in teaching significant math. Obviously there must be an interplay of seriousness and frivolity. The frivolity keeps the reader alert. The seriousness makes the play worthwhile.

That is the kind of mix I have tried to give in my *Scientific American* columns since I started writing them in December, 1956. Six book collections of these columns have previously been published. This is the seventh. As in earlier volumes, the columns have been revised and enlarged to bring them up to date and to include valuable feedback from readers.

The topics covered are as varied as the shows, rides, and concessions of a traveling carnival. It is hoped that the reader who strolls down this colorful mathematical midway, whether he is "with it" as a professional mathematician or just a visiting "mark," will enjoy the noisy fun and games. If he does, he may be surprised, when he finally leaves the lot, by the amount of nontrivial mathematics he has absorbed without even trying.

MARTIN GARDNER
April, 1975

Mathematical Carnival

CHAPTER 1

Sprouts and Brussels Sprouts

I made sprouts fontaneously. . . .
—JAMES JOYCE,
Finnegans Wake, page 542

"A FRIEND of mine, a classics student at Cambridge, introduced me recently to a game called 'sprouts' which became a craze at Cambridge last term. The game has a curious topological flavor."

So began a letter I received in 1967 from David Hartshorne, a mathematics student at the University of Leeds. Soon other British readers were writing to me about this amusing pencil-and-paper game that had sprouted suddenly on the Cambridge grounds.

I am pleased to report that I successfully traced the origin of this game to its source: the joint creative efforts of John Horton Conway, then professor of mathematics at Sidney Sussex College, Cambridge, and Michael Stewart Paterson, then a graduate student working at Cambridge on abstract computer programming theory.

The game begins with n spots on a sheet of paper. Even with as few as three spots, sprouts is more difficult to analyze than ticktacktoe, so that it is best for beginners to play with no more than three or four initial spots. A move consists of drawing a line that joins one spot to another or to itself and then placing a new spot anywhere along the line. These restrictions must be observed:

1. The line may have any shape but it must not cross itself, cross a previously drawn line or pass through a previously made spot.
2. No spot may have more than three lines emanating from it.

Players take turns drawing curves. In normal sprouts, the recommended form of play, the winner is the last person able to play. As in nim and other games of the "take-away" type, the game can also be played in "misère" form, a French term that applies to a variety of card games in the whist family in which one tries to *avoid* taking tricks. In misère sprouts the first person unable to play is the winner.

The typical three-spot normal game shown in Figure 1 was won on the seventh move by the first player. It is easy to see how the game got its name, for it sprouts into fantastic patterns as the game progresses. The most delightful feature is that it is not merely a combinatorial game, as so many connect-the-dots games are, but one that actually exploits the topological properties of the plane. In more technical language, it makes use of the Jordan-curve theorem, which asserts that simple closed curves divide the plane into outside and inside regions.

One might guess at first that a sprouts game could keep sprouting forever, but Conway offers a simple proof that it must end in at most $3n - 1$ moves. Each spot has three "lives"—the three lines that may meet at that point. A spot that acquires three lines is called a "dead spot" because no more lines can be drawn to it. A game that begins with n spots has a starting life of $3n$. Each move kills two lives, at the beginning and at the end of the curve, but adds a new spot with a life of 1. Each move therefore decreases the total life of the game by 1. A game obviously cannot continue when only one life remains, since it requires at least two lives to make a move. Accordingly no game

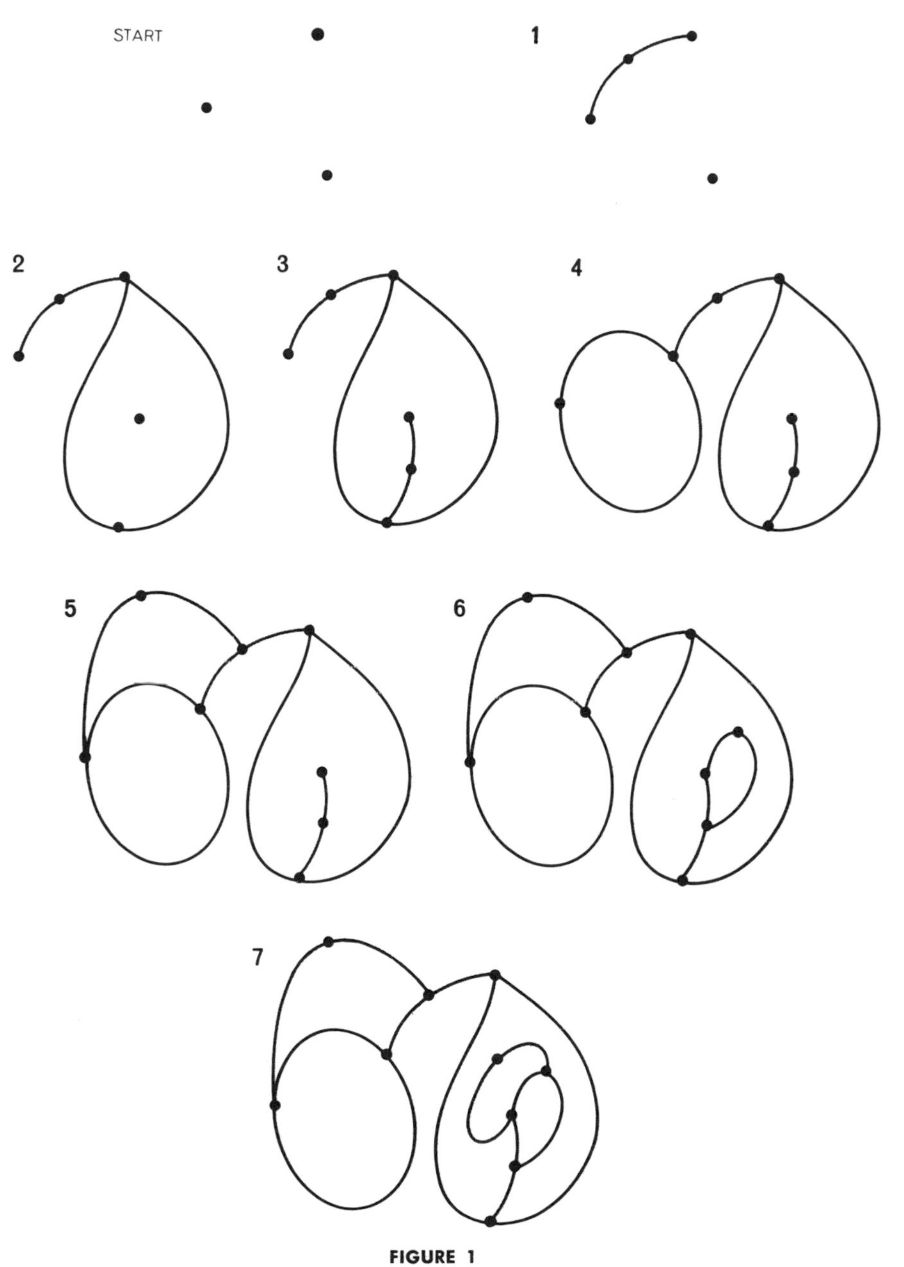

FIGURE 1
A typical game of three-spot sprouts

can last beyond $3n - 1$ moves. It is also easy to show that every game must last at least $2n$ moves. The three-spot game starts with nine lives, must end on or before the eighth move and must last at least six moves.

The one-spot game is trivial. The first player has only one possible move: connecting the spot to itself. The second player wins in the normal game (loses in misère) by joining the two spots, either inside or outside the closed curve. These two second moves are equivalent, as far as playing the game is concerned, because before they are made there is nothing to distinguish the inside from the outside of the closed curve. Think of the game as being played on the surface of a sphere. If we puncture the surface by a hole inside a closed curve, we can stretch the surface into a plane so that all points previously outside the curve become inside, and vice versa. This topological equivalence of inside and outside is important to bear in mind because it greatly simplifies the analysis of games that begin with more than two spots.

With two initial spots, sprouts immediately takes on interest. The first player seems to have a choice of five opening moves [*see Figure 2*], but the second and third openings are equivalent for reasons of symmetry, the same holds true of the fourth and fifth, and in light of the inside-outside equivalence just explained, all four of these moves can be considered identical. Only two topologically distinct moves, therefore, require exploring. It is not difficult to diagram a complete tree chart of all possible moves, inspection of which shows that in both normal and misère forms of the two-spot game the second player can always win.

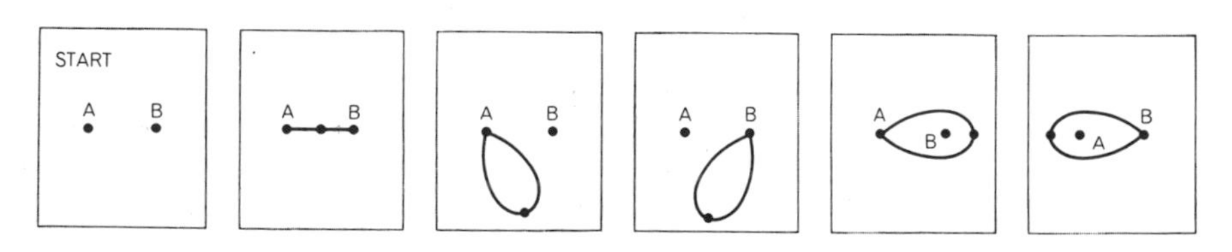

FIGURE 2
Initial spots (A and B) and first player's possible opening moves in two-spot game

Conway found that the first player can always win the normal three-spot game and the second player can always win the misère version. Denis P. Mollison, a Cambridge mathematics student, has shown that the first player has the win in normal four- and five-spot games. In response to a 10-shilling bet made with Conway that he could not complete his analysis within a month, Mollison produced a 49-page proof that the second player wins the normal form of the six-spot game. The second player wins the misère four-spot game. No one yet knows who has the win in misère games that start with more than four spots. Work has been done on the normal game with seven and eight spots, but I know of no results that have been verified. Nor has anyone, to my knowledge, written a satisfactory computer program for analyzing sprouts.

Although no strategy for perfect play has been formulated, one can often see toward the end of a game how to draw closed curves that will divide the plane into regions in such a way as to lead to a win. It is the possibility of this kind of planning that makes sprouts an intellectual challenge and enables a player to improve his skill at the game. But sprouts is filled with unexpected growth patterns, and there seems to be no general strategy that one can adopt to make sure of winning. Conway estimates that a complete analysis of the eight-spot game is beyond the reach of present-day computers.

Sprouts was invented on the afternoon of Tuesday, February 21, 1967, when Conway and Paterson had finished having tea in the mathematics department's common room and were doodling on paper in an effort to devise a new pencil-and-paper game. Conway had been working on a game invented by Paterson that originally involved the folding of attached stamps, and Paterson had put it into pencil-and-paper form. They were thinking of various ways of modifying the rules when Paterson remarked, "Why not put a new dot on the line?"

"As soon as this rule was adopted," Conway has written me, "all the other rules were discarded, the starting position was simplified to just *n* points and sprouts sprouted." The importance of adding the new spot was so great that all parties concerned agree that credit for the game should be on a basis of 3/5 to Paterson and 2/5 to Conway. "And there are complicated rules," Conway adds, "by which we intend to share any monies which might accrue from the game."

"The day after sprouts sprouted," Conway continues, "it seemed that everyone was playing it. At coffee or tea times there were little groups of people peering over ridiculous to fantastic sprout positions. Some people were already attacking sprouts on toruses, Klein bottles, and the like, while at least one man was thinking of higher-dimensional versions. The secretarial staff was not immune; one found the remains of sprout games in the most unlikely places. Whenever I try to acquaint somebody new to the game nowadays, it seems he's already heard of it by some devious route. Even my three- and four-year-old daughters play it, though I can usually beat them."

The name "sprouts" was given the game by Conway. An alternative name, "measles," was proposed by a graduate student because the game is catching and it breaks out in spots, but sprouts was the name by which it quickly became known. Conway later invented a superficially similar game that he calls "Brussels sprouts" to suggest that it is a joke. I shall describe this game but leave to the reader the fun of discovering why it is a joke before the explanation is given in the answer section.

Brussels sprouts begins with n crosses instead of spots. A move consists of extending any arm of any cross into a curve that ends at the free arm of any other cross or the same cross; then a crossbar is drawn anywhere along the curve to create a new cross. Two arms of the new cross will, of course, be dead, since no arm may be used twice. As in sprouts, no curve may cross itself or cross a previously drawn curve, nor may it go through a previously made cross. As in sprouts, the winner of the normal game is the last person to play and the winner of the misère game is the first person who cannot play.

After playing sprouts, Brussels sprouts seems at first to be a more complicated and more sophisticated version. Since each move kills two crossarms and adds two live crossarms, presumably a game might never end. Nevertheless, all games do end and there is a concealed joke that the reader will discover if he succeeds in analyzing the game. To make the rules clear, a typical normal game of two-cross Brussels sprouts is shown that ends with victory for the second player on the eighth move [*see Figure 3*].

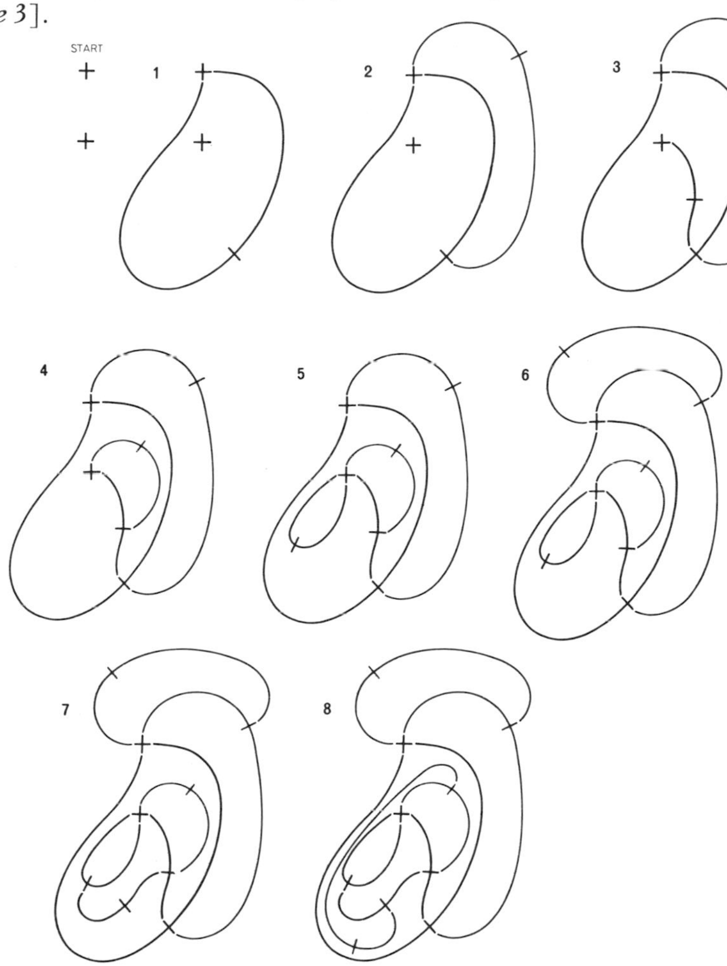

FIGURE 3
Typical game of two-cross Brussels sprouts

A letter from Conway reports several important breakthroughs in sproutology. They involve a concept he calls the "order of moribundity" of a terminal position, and the classification of "zero order" positions into five basic types: louse, beetle, cockroach, earwig, and scorpion. The larger insects and arachnids can be infested with lice, sometimes in nested form, and Conway draws one pattern he says is "merely an inside-out earwig inside an inside-out louse." Certain patterns, he points out, are much lousier than others. And there is the FTOZOM (fundamental theorem of zero-order moribundity), which is quite deep. Sproutology is sprouting so rapidly that I shall have to postpone my next report on it for some time.

ADDENDUM

SPROUTS MADE an instant hit with *Scientific American* readers, many of whom suggested generalizations and variations of the game. Ralph J. Ryan III proposed replacing each spot with a tiny arrow, extending from one side of the line, and allowing new lines to be drawn only to the arrow's point. Gilbert W. Kessler combined spots and crossbars in a game he called "succotash." George P. Richardson investigated sprouts on the torus and other surfaces. Eric L. Gans considered a generalization of Brussels sprouts (called "Belgian sprouts") in which spots are replaced by "stars"—n crossbars crossing at the same point. Vladimir Ygnetovich suggested the rule that a player, on each turn, has a choice of adding one, two, or no spots to his line.

Several readers questioned the assertion that every game of normal sprouts must last at least $2n$ moves. They sent what they believed to be counterexamples, but in each case failed to notice that every isolated spot permits two additional moves.

ANSWERS

WHY IS the game of Brussels sprouts, which appears to be a more sophisticated version of sprouts, considered a joke by its inventor, John Horton Conway? The answer is that it is impossible to play Brussels sprouts either well or poorly because every game must end in exactly $5n - 2$ moves, where n is the number of initial crosses. If played in standard form (the last to play is the winner), the game is always won by the first player if it starts with an odd number of crosses, by the second player if it starts with an even number. (The reverse is true, of course, in misère play.) After introducing someone to sprouts, which is a genuine contest, one can switch to the fake game of Brussels sprouts, make bets and always know in advance who will win every game. I leave to the reader the task of proving that each game must end in $5n - 2$ moves.

CHAPTER 2

Penny Puzzles

Coins have a variety of simple properties that can be exploited in recreational mathematics: they stack easily, they can be used as counters, they can serve as models for points on the plane, they are circular, and they have two distinguishable sides. Here is a collection of entertaining coin puzzles that require no more than 10 pennies. They are elementary enough to make excellent bar or dinner-table diversions, yet some of them lead into areas of mathematics that are far from trivial.

One of the oldest and best coin puzzles calls for placing eight pennies in a row on the table [*see Figure 4*] and trying to transform them, in four moves, into four stacks of two coins each. There is one proviso: on each move a single penny must "jump" exactly two pennies (in either direction) and land on the next single penny. The two jumped pennies may be two single coins side by side or a stacked pair. Eight is the smallest number of pennies that can be paired in this manner.

FIGURE 4
The doubling problem

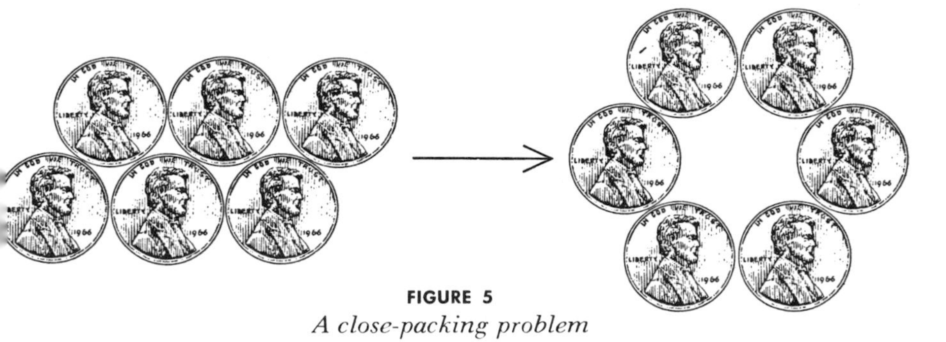

FIGURE 5
A close-packing problem

The reader will find it a pleasant and easy task to solve the problem, but now comes the amusing part. Suppose two more coins are added to make a row of 10? Can the 10 be doubled in five moves? Many people, after finding a solution for eight, will give up in despair when presented with 10 and refuse to work on the problem. Yet it can be solved instantly if one has the right insight. Indeed, as will be explained in the answer section, a solution for eight makes trivial the generalization to a row of $2n$ pennies ($n > 3$), to be doubled in n moves.

When pennies are closely packed on the plane, their centers mark the points of a triangular lattice, a fact that underlies scores of coin puzzles of many different types. For example, start with six pennies closely packed in a rhombus formation [*at left in Figure* 5]. In three moves try to form a circular pattern [*at right in Figure* 5] so that if a seventh coin were placed in the center, the six pennies would be closely packed around it. On each move a single coin must be slid to a new position so as to touch two other coins that rigidly determine its new position. Like the doubling puzzle, this too has a tricky aspect when you show it to someone. If he fails to solve it, demonstrate the solution slowly and challenge him to repeat it. But when you put the coins back in their starting position, set them in the mirror-image form of the original rhombus. There is a good chance that he will not notice the difference, with the result that when he tries to duplicate your three moves, he will soon be in serious trouble.

A good follow-up to the preceding puzzle is to pack 10 pennies to form a triangle [*at left in Figure 6*]. This is the famous "tetractys" of the ancient Pythagoreans and today's familiar pattern for the 10 bowling pins. The problem is to turn this triangle upside down [*at right in Figure 6*] by sliding one penny at a time, as before, to a new position in which it touches two other pennies. What is the minimum number of required moves? Most people solve the problem quickly in four moves, but it can be done in three. The problem has an interesting generalization. A triangle of three pennies obviously can be inverted by moving one coin and a triangle of six pennies by moving two. Since the 10-coin triangle calls for a shift of three, can the next largest equilateral triangle, 15 pennies arranged like the 15 balls at the start of a billiard game, be inverted by moving four pennies? No, it requires five. Nevertheless, there is a remarkably simple way to calculate the minimum number of coins that must be shifted, given the number of pennies in the triangle. Can the reader discover it?

The tetractys also lends itself to a delightful puzzle of the peg-solitaire type. Peg solitaire, traditionally played on a square lattice, is an old recreation that is now the topic of a considerable

FIGURE 6
Triangle-inversion problem

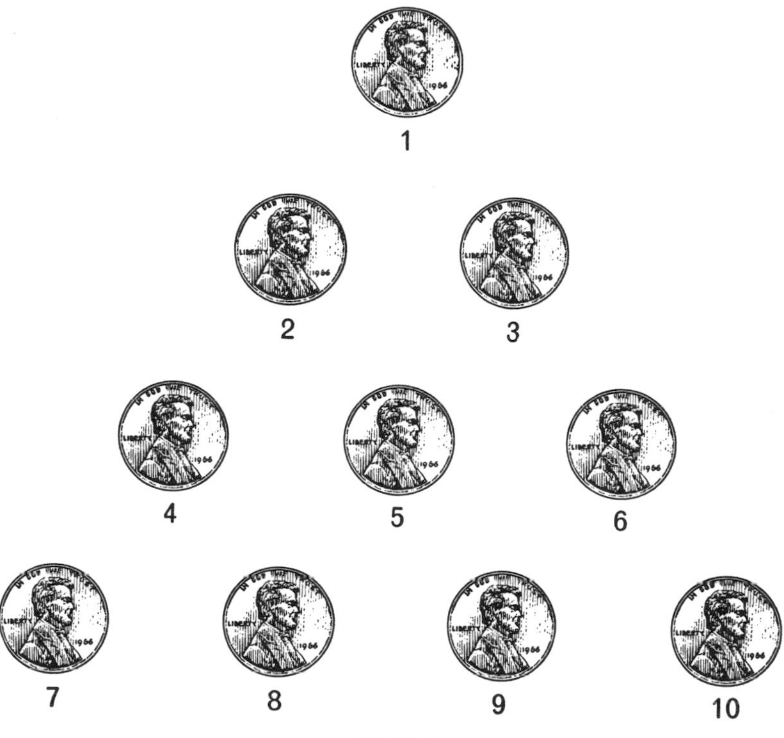

FIGURE 7
Triangular solitaire

literature. As far as I know, similar problems based on the triangular lattice have received only the most superficial attention. The simplest starting position that is not trivial is the Pythagorean 10-spot triangle. To make it easy to record solutions, draw 10 spots on a sheet of paper, spacing them so that when pennies are placed on the spots, there will be some space between them, and number the positions [*see Figure 7*].

The problem: Remove one penny to make a "hole," then reduce the coins to a single penny by jumping. No sliding moves are allowed. The jumps are as in checkers, over an adjacent coin to an empty spot immediately beyond. The jumped coin is removed. Note that jumps can be made in six directions: in either direction parallel with the base, and in either direction parallel with each of the triangle's other two sides. As in checkers, a continuous chain of jumps counts as one "move." After some trial-and-error tests one discovers that the puzzle is indeed solvable, but of course the recreational mathematician is not happy until he solves it in a minimum number of moves. Here, for example, is a solution in six moves after removal of the coin on spot 2:

1. 7–2
2. 9–7
3. 1–4
4. 7–2
5. 6–4, 4–1, 1–6
6. 10–3

There is a better solution in five moves. Can the reader find it? If so, he may wish to go on to the 15-spot triangle. The novelty company S. S. Adams has been selling a peg version of this for years under the trade name Ke Puzzle Game, but no solutions are provided with the puzzle.

If one penny rolls around another penny without slipping, how many times will it rotate in making one revolution? One might guess the answer to be one, since the moving penny rolls along an edge equal to its own circumference, but a quick experiment shows that the answer is two; apparently the complete revolution of the moving penny adds an extra rotation. Suppose we roll the penny, without slipping, from the top of a six-penny triangle [*see Figure 8*] all the way around the sides and back to the starting position. How many times will it rotate? It is easy to see from the picture that the penny rolls along arcs with a total distance (expressed in fractions of full circles) of $^{12}/_{6}$, or two full circles. Therefore it must rotate at least twice. Because it makes a complete revolution, shall we add one more rotation and say it rotates three times? No, a test discloses that it makes four rotations! The truth is that for every degree of arc along which it rolls, it rotates two degrees. We must double the length of the path along which it rolls to get the correct answer: four rotations. With this in mind it is easy to solve other puzzles of this type, which are often found in puzzle books. Simply calcu-

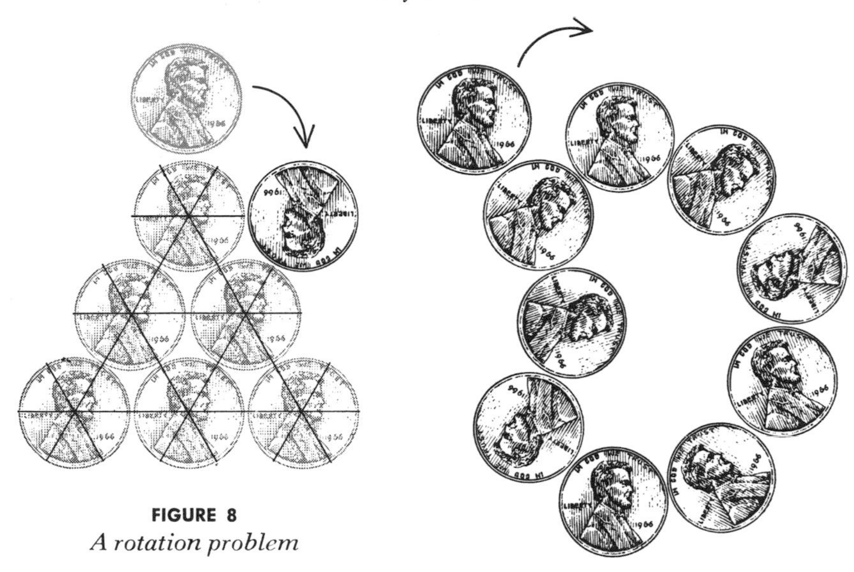

FIGURE 8
A rotation problem

FIGURE 9
A surprising invariance theorem

late the length of the path in degrees, multiply by two, and you have the number of degrees of rotation.

All of this is fairly obvious, but there is a beautiful theorem lurking here that has not been noticed before to my knowledge. Instead of close-packing the coins around which a coin is rolled, join them to form an irregular closed chain; Figure 9 shows a random chain of nine pennies. (The only proviso is that as a penny rolls around it without slipping, the rolling coin must touch every coin in the chain.) It turns out, surprisingly, that regardless of the shape of a chain of a given length, the number of rotations made by the rolling penny by the time it returns to its starting position is always the same! In the nine-penny case the penny makes exactly five rotations. If the penny is rolled *inside* the chain, it will make exactly one rotation. This is also a constant, unaffected by the shape of the chain. Can the reader prove (only the most elementary geometry need be used) that for any closed chain of n pennies ($n > 2$) the number of rotations of the moving penny, as it rolls once around the outside of the chain, is constant? If so, he will see immediately how to ap-

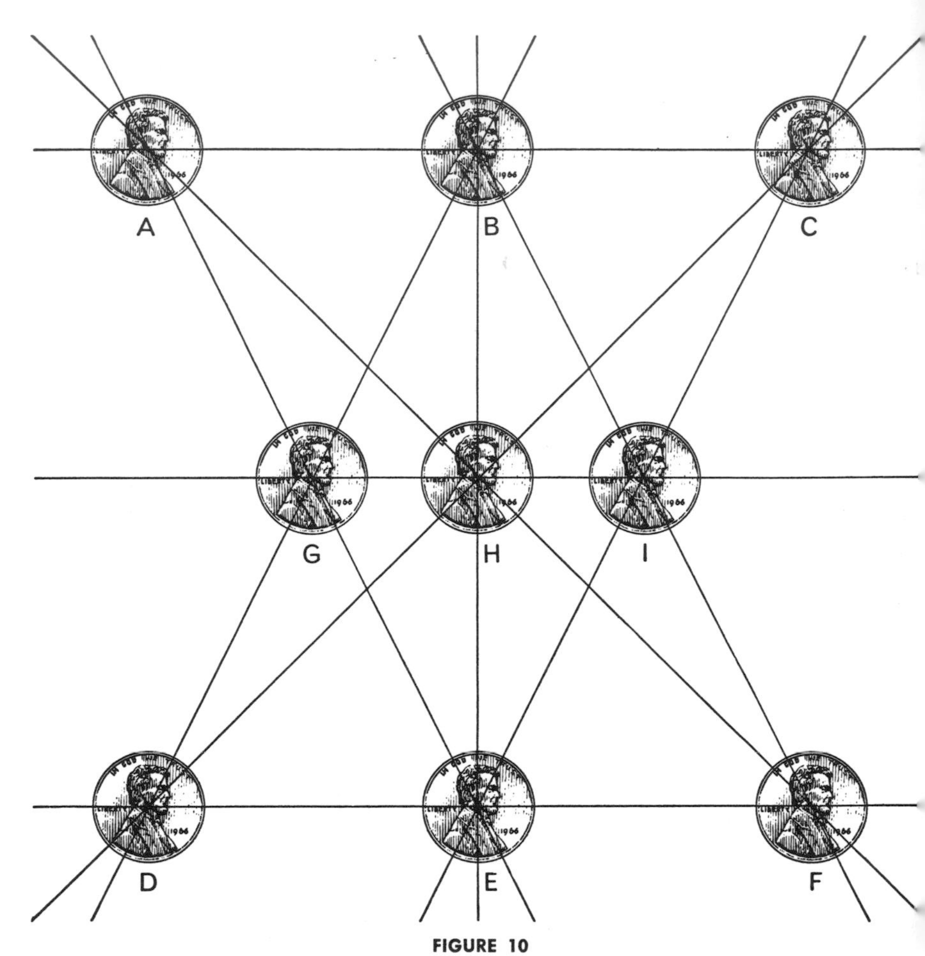

FIGURE 10
Tree-planting problem and Pappus' theorem

ply the same proof to a penny rolling inside a chain of n coins ($n > 6$) as well as how to derive a simple formula, for each case, that expresses the number of rotations as a function of n.

Pennies are also convenient markers for working on what in the puzzle field are called "tree-planting problems." For example, a farmer wishes to plant nine trees so that they form 10 straight rows with three trees in each row. If the reader is familiar with projective geometry, he may notice that the solution [*see Figure 10*] could be taken as a diagram for the famous theorem of Pappus: If three points A, B, C are located anywhere along one line, and three points D, E, F are located anywhere on a second line (the two lines need not be parallel as here), the intersections *G*, *H*, *I* of opposite sides of the crossed hexagon *AFBDCE* will be on a straight line. The Pappus theorem therefore guarantees nine three-point lines; the 10th line is added by adjusting the figure to bring points *B*, *H*, *E* into line.

Tree-planting puzzles tie in with an aspect of projective geometry called "incidence geometry" (a point is incident to any line passing through it, and a line is incident to any point it passes through). Harold L. Dorwart of Trinity College, Hartford, Conn., has published a splendid popular introduction to this subject, *The Geometry of Incidence* (Prentice-Hall), which I recommend. On page 146 he cites two tree-planting problems —25 trees in 10 rows of six to a row and 19 trees in nine rows of five to a row—both of which are solved by inspecting the figure for the famous projective theorem of Desargues. Tree-planting problems lead into deep combinatorial waters. No one has yet discovered a general procedure by which all problems of this kind can be solved, and the field is therefore riddled with unanswered questions.

Now back to our pennies. It turns out that 10 of them can be arranged to form five straight lines, with four in each line. (It is assumed, of course, that each line must pass through the centers of the four coins on it.) Figure 11 shows five ways it can be done. Each pattern can be distorted in an infinity of ways without changing its topological structure; they are shown here in forms given by the English puzzlist Henry Ernest Dudeney to display bilateral symmetry for all of them. There is a sixth solution, topologically distinct from the other five. Can the reader discover it?

Many coin puzzles combine a bit of mathematics with "catch" features that make them excellent quickies of the "bar-bet" type. For example, arrange four pennies in a square formation on the bar and bet someone you can change the position of only one penny and produce two straight rows with three pennies in each row. It looks impossible, but the solution is to pick up a penny and put it on *top* of the penny diagonally opposite.

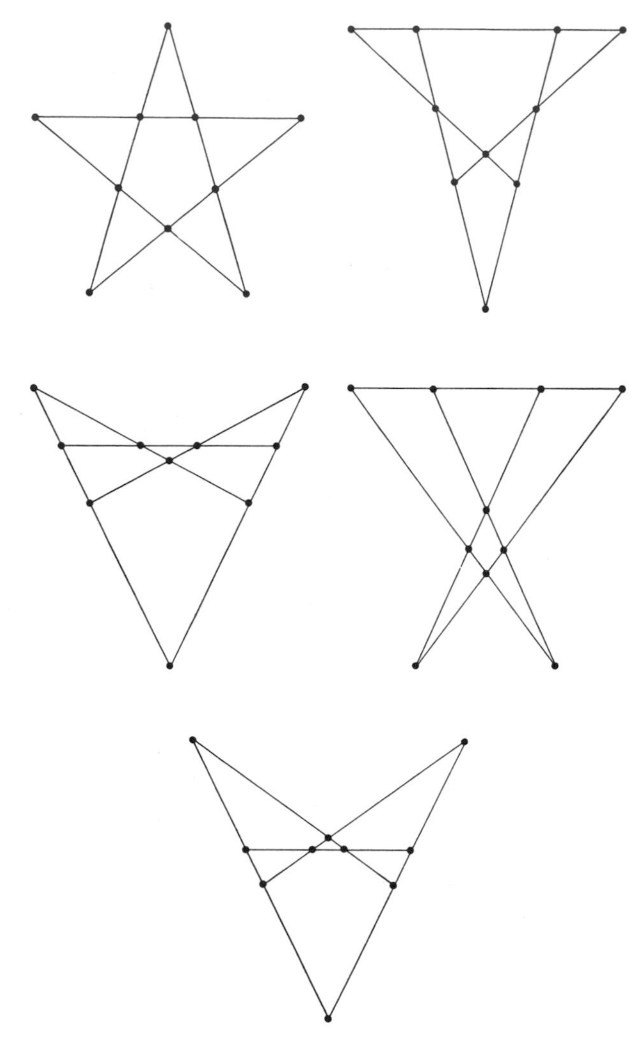

FIGURE 11
Five ways to plant 10 trees in five rows of four each

Here are two more penny catches: First, arrange three pennies as shown in Figure 12 and challenge someone to put penny *C* between *A* and *B* so that the three coins lie in a straight line—without moving *B* in any way, without touching *A* with his hands, with any part of his body, or with any object, and

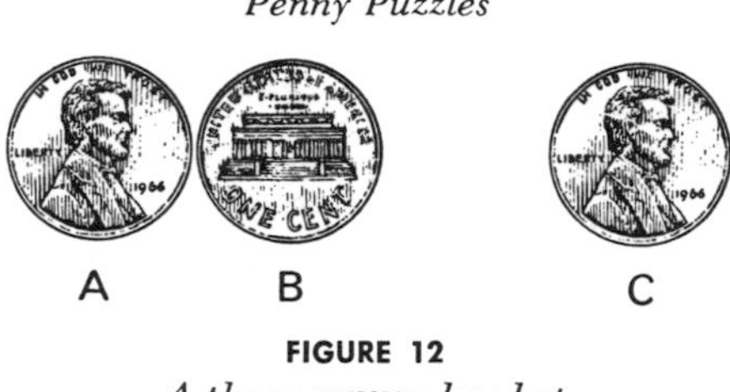

FIGURE 12
A three-penny bar bet

without moving *A* by blowing on it. For yet another bet, draw a vertical line on a sheet of paper. The challenge is to position three pennies in such a way that the surfaces of two heads are wholly to the right of the line and the surfaces of two tails are wholly to the left of the line.

ADDENDUM

TRIANGULAR SOLITAIRE obviously can be played on triangular matrices with borders other than the triangle: hexagons, rhombuses, six-pointed stars, and so on. It also can be played with varying rules such as: (1) Jumps parallel to one side of the unit cell can be prohibited. See the 15-triangle solitaire problem in Maxey Brooke, *Fun for the Money* (Scribner's, 1963), page 12. (2) Sliding moves, as well as jumps, may be allowed, as in the game of Chinese checkers.

If isometric (triangular) solitaire is played in the classic way, allowing jumps only, and in all six directions, there are elegant procedures for testing whether a given pattern can be derived from another. The procedures are extensions of those which have been developed for square solitaire. (See the chapter on peg solitaire in my *Unexpected Hanging and Other Mathematical Diversions*, Simon & Schuster, 1969.)

As in square solitaire, the procedures do not provide actual solutions, nor do they prove that a solution exists, but they are capable of showing certain problems to be unsolvable. Much unpublished work on this has been done by Mannis Charosh, Harry O. Davis, John Harris, and Wade E. Philpott. The methods are all based on a commutative group which determines a pattern's parity. By coloring the lattice points with three colors, and following various rules, one can quickly determine the impossibility of certain problems. For example, a hexagonal field with a center vacancy cannot be reduced to one counter in the center unless the side of the hexagon is a number of the form $3n + 2$. A clear explanation of one such procedure is given by Irvin Roy Hentzel in "Triangular Puzzle Peg," *Journal of Recreational Mathematics*, Vol. 6, Fall, 1973, pages 280–283.

ANSWERS

1. To DOUBLE the eight pennies in a row into four stacks of two coins each, number them from one to eight and move 4 to 7, 6 to 2, 1 to 3, 5 to 8. For 10 pennies, simply double the pennies at one end (for example, move 7 to 10), to leave a row of eight that can be solved as before. Clearly, a row of $2n$ pennies can be solved in n moves by doubling the coins at one end until eight remain, then solving for the eight.

2. Six pennies in rhombus formation [*Figure 13*] can be formed into a circle, in accordance with the rules given, by numbering them as shown and moving 6 to touch 4 and 5, 5 to touch 2 and 3 from below, 3 to touch 5 and 6.

FIGURE 13
Rhombus-to-circle puzzle

FIGURE 14
Inverting the triangle

3. A triangle of 10 pennies is inverted by moving three coins as shown in Figure 14. In working on the general problem, for equilateral formations of any size, readers may have realized that the problem is one of drawing a bounding triangle (like the frame used to group the 15 balls for a game of billiards), inverting it and placing it over the figure so that it encloses a maximum number of coins. In every case the smallest number of coins that must be moved to invert the pattern is obtained by dividing the number of coins by 3 and ignoring the remainder.

4. The 10-coin triangle, with positions numbered as shown in Figure 7, is reduced to one coin in five moves by first removing the penny at spot 3, then jumping: 10–3, 1–6, 8–10–3, 4–6, 1–4, 7–2. The solution is unique except for the fact that the triple jump can be made clockwise or counterclockwise. Of course, the initial vacancy may be at any of the six spots not at a corner or in the center.

Nine moves are minimal for the 15-coin triangle. The vacancy must be in the middle of a side, and the first two moves must be 1–4, 7–2, or one of the five other pairs which are symmetrically equivalent. With the first two moves specified, a computer program by Malcolm E. Gillis, Jr., of Slidell, La., found 260 solutions.

The following solution is one of many that end with a dramatic five-jump sweep. The positions are numbered left to right, top to bottom: (1) 11–4, (2) 2–7, (3) 13–4, (4) 7–2, (5) 15–13, (6) 12–14, (7) 10–8, (8) 3–10, (9) 1–4, 4–13, 13–15, 15–6, 6–4.

I know of no computer analysis of the 21-triangle or any triangle of higher order. John Harris, of Santa Barbara, Cal., proved that nine moves are necessary for the 21-triangle, and the following solution, from Edouard Marmier, Zurich, proves that nine moves are also sufficient: (1) 1–4, (2) 7–2, (3) 16–7, (4) 6–1, 1–4, 4–11, (5) 13–6, 6–4, 4–13, (6) 18–16, 16–7, 7–18, 18–9, (7) 15–6, 6–13, (8) 20–18, 18–9, 9–20, (9) 21–19.

5. Readers were asked to prove that if a penny is rolled once around a closed chain of pennies, touching every penny, the number of rotations made by the penny is a constant regardless of the chain's shape. We shall prove this first for a chain of nine pennies.

Join the centers of the coins by straight lines as shown at the left of Figure 15 to form a nine-sided polygon. The total length (in degrees of arc) of the perimeters of the pennies, outside the polygon, is the same as the total of the polygon's conjugate angles. (The conjugate of an angle is the difference between that angle and 360 degrees.) The sum of the conjugate angles of a polygon of n sides is always $\frac{1}{2}n + 1$ perigons (a perigon is an angle of 360 degrees).

As the penny rolls around the chain, however, for every pair of pennies it touches, it fails to touch two arcs of $\frac{1}{6}$ perigon each, which together are $\frac{1}{3}$ perigon [*see right of Figure 15*]. For n pennies, it will fail to touch $n/3$ perigons. We subtract this from $\frac{1}{2}n + 1$ to obtain $\frac{1}{6}n + 1$ as the total perimeter over which the penny rolls in making one circuit around the chain.

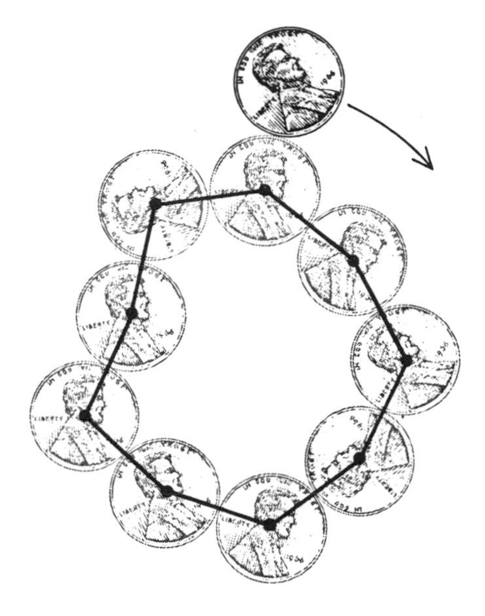

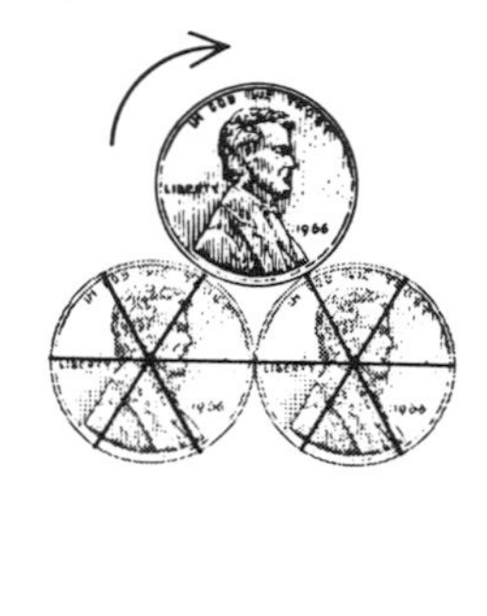

FIGURE 15
The rolling penny

As explained, the penny rotates two degrees for each degree of arc over which it rolls, therefore the penny must make a total of $\frac{1}{3}n + 2$ rotations. This obviously is a constant, regardless of the number of pennies or the chain's shape, because the centers of any chain of pennies must mark the vertices of an n-sided polygon. (The formula also applies to a degenerate chain of two pennies, whose centers can be taken as the corners of a degenerate polygon of two sides.)

A similar argument establishes $(n/3) - 2$ as the number of rotations made by a penny rolling once around the inside of a closed chain of six or more pennies and touching every penny. The formula gives zero rotations for the chain of six, inside which it fits snugly, touching all six coins. For an open chain of n pennies it is easy to show that the rolling penny makes $\frac{1}{3}(2n + 4)$ rotations in a complete circuit.

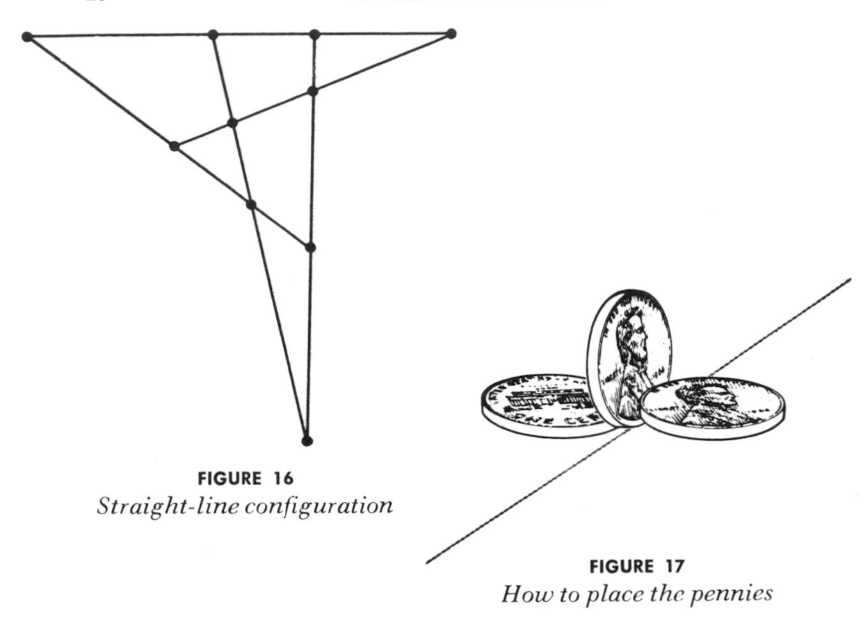

FIGURE 16
Straight-line configuration

FIGURE 17
How to place the pennies

6. The sixth configuration of 10 pennies having five straight lines of four coins each is shown in Figure 16.

7. To put penny C between two touching pennies A and B without touching A or moving B, place a fingertip firmly on B and then slide C against B. Be sure, however, to let go of C before it strikes B. The impact will propel A away from B, so that C can be placed between the two previously touching coins.

8. Three pennies can be placed with two heads on one side of a line and two tails on the other as shown in Figure 17.

CHAPTER 3

Aleph-null and Aleph-one

A graduate student at Trinity
Computed the square of infinity.
But it gave him the fidgets
To put down the digits,
So he dropped math and took up divinity
—ANONYMOUS

IN 1963 Paul J. Cohen, a 29-year-old mathematician at Stanford University, found a surprising answer to one of the great problems of modern set theory: Is there an order of infinity higher than the number of integers but lower than the number of points on a line? To make clear exactly what Cohen proved, something must first be said about those two lowest known levels of infinity.

It was Georg Ferdinand Ludwig Philipp Cantor who first discovered that beyond the infinity of the integers—an infinity to which he gave the name aleph-null—there are not only higher infinities but also an infinite number of them. Leading mathematicians were sharply divided in their reactions. Henri Poincaré called Cantorism a disease from which mathematics would have to recover, and Hermann Weyl spoke of Cantor's hierarchy of alephs as "fog on fog."

On the other hand, David Hilbert said, "From the paradise

created for us by Cantor, no one will drive us out," and Bertrand Russell once praised Cantor's achievement as "probably the greatest of which the age can boast." Today only mathematicians of the intuitionist school and a few philosophers are still uneasy about the alephs. Most mathematicians long ago lost their fear of them, and the proofs by which Cantor established his "terrible dynasties" (as they have been called by the world-renowned Argentine writer Jorge Luis Borges) are now universally honored as being among the most brilliant and beautiful in the history of mathematics.

Any infinite set of things that can be counted 1, 2, 3 . . . has the cardinal number $\aleph_0$ (aleph-null), the bottom rung of Cantor's aleph ladder. Of course, it is not possible actually to count such a set; one merely shows how it can be put into one-to-one correspondence with the counting numbers. Consider, for example, the infinite set of primes. It is easily put in one-to-one correspondence with the positive integers:

$$\begin{array}{cccccc} 1 & 2 & 3 & 4 & 5 & 6\ldots \\ \downarrow & \downarrow & \downarrow & \downarrow & \downarrow & \downarrow \\ 2 & 3 & 5 & 7 & 11 & 13\ldots \end{array}$$

The set of primes is therefore an aleph-null set. It is said to be "countable" or "denumerable." Here we encounter a basic paradox of all infinite sets. Unlike finite sets, they can be put in one-to-one correspondence with a *part* of themselves or, more technically, with one of their "proper subsets." Although the primes are only a small portion of the positive integers, as a completed set they have the same aleph number. Similarly, the integers are only a small portion of the rational numbers (the integers plus all integral fractions), but the rationals form an aleph-null set too.

There are all kinds of ways in which this can be proved by arranging the rationals in a countable order. The most familiar way is to attach them, as fractions, to an infinite square array of lattice points and then count the points by following a zigzag path, or a spiral path if the lattice includes the negative rationals. Here is another method of ordering and counting the positive rationals that was proposed by the American logician Charles Sanders Peirce. (See *Collected Papers of Charles Sanders Peirce,* Harvard University Press, 1933, pages 578–580.)

Start with the fractions 0/1 and 1/0. (The second fraction is meaningless, but that can be ignored.) Sum the two numerators and then the two denominators to get the new fraction 1/1, and place it between the previous pair: 0/1, 1/1, 1/0. Repeat this procedure with each pair of adjacent fractions to obtain two new fractions that go between them:

$$\frac{0}{1} \quad \frac{1}{2} \quad \frac{1}{1} \quad \frac{2}{1} \quad \frac{1}{0}.$$

The five fractions grow, by the same procedure, to nine:

$$\frac{0}{1} \quad \frac{1}{3} \quad \frac{1}{2} \quad \frac{2}{3} \quad \frac{1}{1} \quad \frac{3}{2} \quad \frac{2}{1} \quad \frac{3}{1} \quad \frac{1}{0}.$$

In this continued series every rational number will appear once and only once, and always in its simplest fractional form. There is no need, as there is in other methods of ordering the rationals, to eliminate fractions, such as 10/20, that are equivalent to simpler fractions also on the list, because no reducible fraction ever appears. If at each step you fill the cracks, so to speak, from left to right, you can count the fractions simply by taking them in their order of appearance.

This series, as Peirce said, has many curious properties. At each new step the digits above the lines, taken from left to right, begin by repeating the top digits of the previous step: 01, 011, 0112, and so on. And at each step the digits below the lines are the same as those above the lines but in reverse order. As a consequence, any two fractions equally distant from the central 1/1 are reciprocals of each other. Note also that for any adjacent pair, a/b, c/d, we can write such equalities as $bc - ad = 1$, and $c/d - a/b = 1/bd$. The series is closely related to what are called Farey numbers (after the English geologist John Farey, who

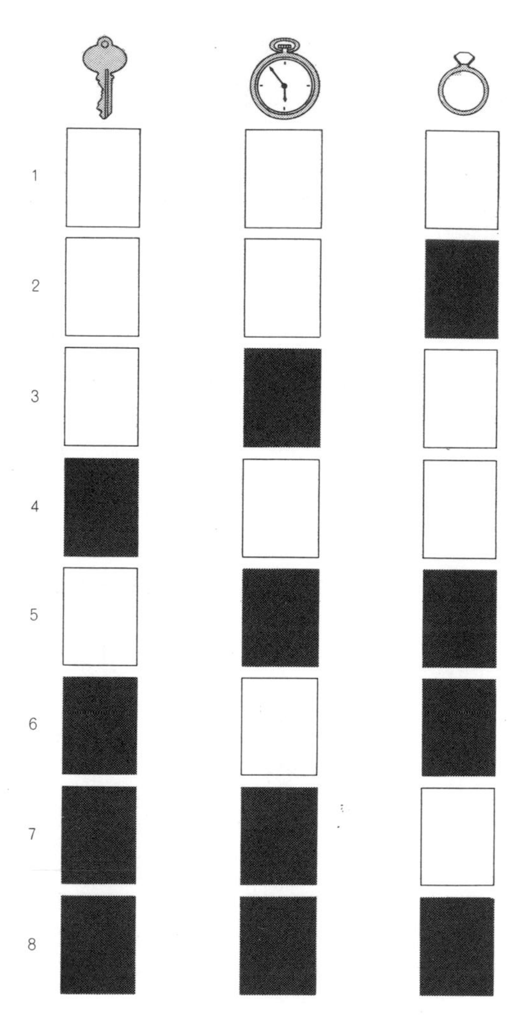

FIGURE 18
Subsets of a set of three elements

first analyzed them), about which there is now a considerable literature.

It is easy to show that there is a set with a higher infinite number of elements than aleph-null. To explain one of the best of such proofs, a deck of cards is useful. First consider a finite set of three objects, say a key, a watch, and a ring. Each subset of this set is symbolized by a row of three cards [*see Figure 18*], a face-up card (white) indicates that the object above it is in the subset, a face-down card (gray) indicates that it is not. The first subset consists of the original set itself. The next three rows indicate subsets that contain only two of the objects. They are followed by the three subsets of single objects and finally by the empty (or null) subset that contains none of the objects. For any set of n elements the number of subsets is 2^n. (It is easy to see why. Each element can be either included or not, so for one element there are two subsets, for two elements there are $2 \times 2 = 4$ subsets, for three elements there are $2 \times 2 \times 2 = 8$ subsets, and so on.) Note that this formula applies even to the empty set, since $2^0 = 1$ and the empty set has the empty set as its sole subset.

This procedure is applied to an infinite but countable (aleph-null) set of elements at the left in Figure 19. Can the subsets of this infinite set be put into one-to-one correspondence with the counting integers? Assume that they can. Symbolize each subset with a row of cards, as before, only now each row continues endlessly to the right. Imagine these infinite rows listed in any order whatever and numbered 1, 2, 3 . . . from the top down.

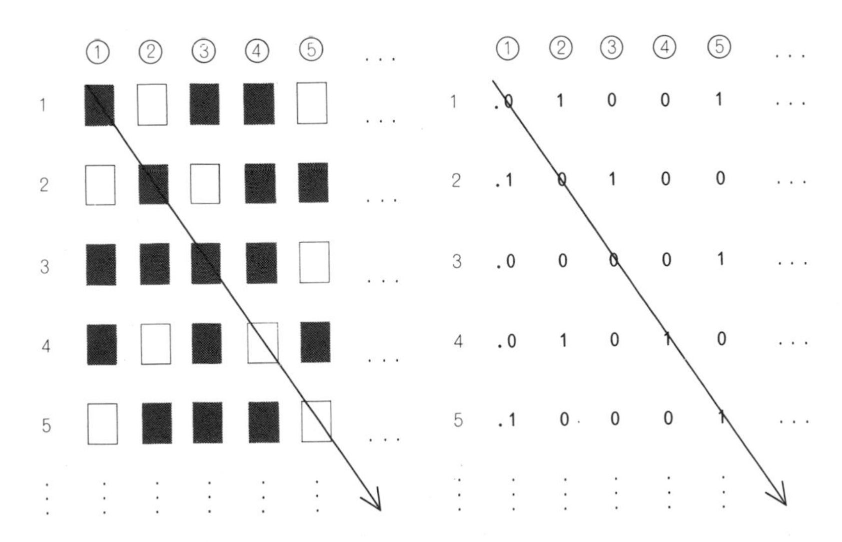

FIGURE 19

A countable infinity has an uncountable infinity of subsets (left) that correspond to the real numbers (right)

If we continue forming such rows, will the list eventually catch all the subsets? No—because there is an infinite number of ways to produce a subset that cannot be on the list. The simplest way is to consider the diagonal set of cards indicated by the arrow and then suppose every card along this diagonal is turned over (that is, every face-down card is turned up, every face-up card is turned down). The new diagonal set cannot be the first subset because its first card differs from the first card of subset 1. It cannot be the second subset because its second card differs from the second card of subset 2. In general it cannot be the *n*th subset because its *n*th card differs from the *n*th card of subset *n*. Since we have produced a subset that cannot be on the list, even when the list is infinite, we are forced to conclude that the original assumption is false. The set of all subsets of an aleph-null set is a set with the cardinal number 2 raised to the power of aleph-null. This proof shows that such a set cannot be matched one to one with the counting integers. It is a higher aleph, an "uncountable" infinity.

Cantor's famous diagonal proof, in the form just given, conceals a startling bonus. It proves that the set of real numbers (the rationals plus the irrationals) is also uncountable. Consider a line segment, its ends numbered 0 and 1. Every rational fraction from 0 to 1 corresponds to a point on this line. Between any two rational points there is an infinity of other rational points; nevertheless, even after all rational points are identified, there remains an infinity of unidentified points—points that correspond to the unrepeating decimal fractions attached to such algebraic irrationals as the square root of 2, and to such transcendental irrationals as pi and *e*. Every point on the line segment, rational or irrational, can be represented by an endless decimal fraction. But these fractions need not be decimal; they can also be written in binary notation. Thus every point on the line segment can be represented by an endless pattern of 1's and 0's, and every possible endless pattern of 1's and 0's corresponds to exactly one point on the line segment.

Now, suppose each face-up card at the left in Figure 19 is replaced by 1 and each face-down card by 0, as shown at the right in the illustration. We have only to put a binary point in front of each row and we have an infinite list of different binary fractions between 0 and 1. But the diagonal set of symbols, after each 1 is changed to 0 and each 0 to 1, is a binary fraction that cannot be on the list. From this we see that there is a one-to-one correspondence of three sets: the subsets of aleph-null, the real numbers (here represented by binary fractions), and the totality of points on a line segment. Cantor gave this higher infinity the cardinal number *C*, for the "power of the continuum." He believed it was also $\aleph_1$ (aleph-one), the first infinity greater than aleph-null.

By a variety of simple, elegant proofs Cantor showed that *C* was the number of such infinite sets as the transcendental irrationals (the algebraic irrationals, he proved, form a countable set), the number of points on a line of infinite length, the number of points on any plane figure or on the infinite plane, and the number of points in any solid figure or in all of three-space. Going into higher dimensions does not increase the number of points. The points on a line segment one inch long can be matched one to one with the points in any higher-dimensional solid, or with the points in the entire space of any higher dimension.

The distinction between aleph-null and aleph-one (we accept, for the moment, Cantor's identification of aleph-one with *C*) is important in geometry whenever infinite sets of figures are encountered. Imagine an infinite plane tessellated with hexagons. Is the total number of vertices aleph-one or aleph-null? The answer is aleph-null; they are easily counted along a spiral path [*see Figure 20*]. On the other hand, the number of different circles of one-inch radius that can be placed on a sheet of typewriter paper is aleph-one because inside any small square near the center of the sheet there are aleph-one points, each the center of a different circle with a one-inch radius.

Consider in turn each of the five symbols J. B. Rhine uses on his "ESP" test cards [*see Figure 21*]. Can it be drawn an aleph-

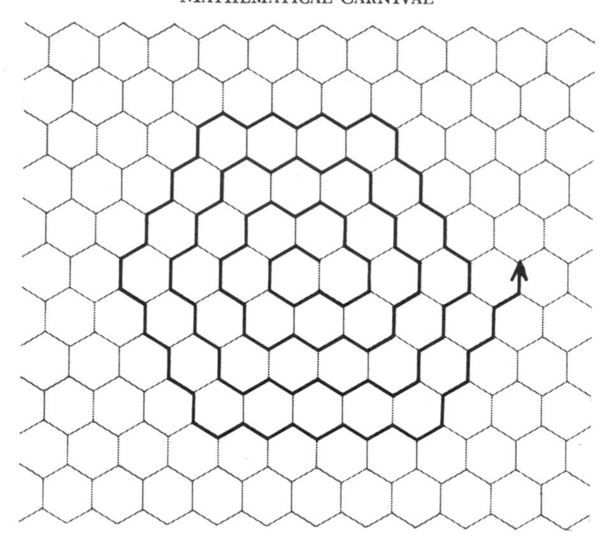

FIGURE 20
Spiral counts the vertices of a hexagonal tessellation

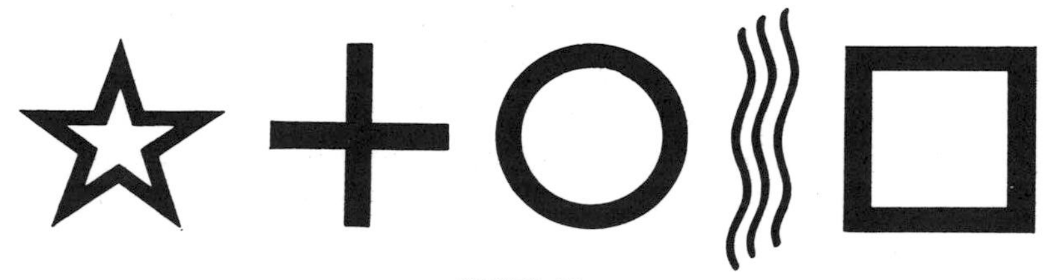

FIGURE 21
Five "ESP" symbols

one number of times on a sheet of paper, assuming that the symbol is drawn with ideal lines of no thickness and that there is no overlap or intersection of any lines? (The drawn symbols need not be the same size, but all must be similar in shape.) It turns out that all except one can be drawn an aleph-one number of times. Can the reader show which symbol is the exception?

The two alephs are also involved in recent cosmological speculation. Richard Schlegel, a physicist, has called attention in several papers to a strange contradiction inherent in the "steady-state" theory. According to that theory, the number of atoms in the cosmos at the present time is aleph-null. (The cosmos is regarded as infinite even though an "optical horizon" puts a limit on what can be seen.) Moreover, atoms are steadily increasing in number as the universe expands. Infinite space can easily accommodate any finite number of doublings of the quantity of atoms, for whenever aleph-null is multiplied by two, the result is aleph-null again. (If you have an aleph-null number of eggs in aleph-null boxes, one egg per box, you can accommodate another aleph-null set of eggs simply by shifting the egg in box 1 to box 2, the egg in box 2 to box 4, and so on, each egg going to a box whose number is twice the number of the egg's previous box. This empties all the odd-numbered boxes, which can then be filled with another aleph-null set of eggs.) But if the doubling goes on for an aleph-null number of times, we come up against the formula of 2 raised to the power of aleph-null—that is, $2 \times 2 \times 2 \ldots$ repeated aleph-null times. As we have seen, this produces an aleph-one set. Consider only two atoms at an infinitely remote time in the past. By now, after an aleph null series of doublings, they would have grown to an aleph-one set. But the cosmos, at the moment, cannot contain an aleph-one set of atoms. Any collection of distinct physical entities (as opposed to the ideal entities of mathematics) is countable and therefore, *at the most,* aleph-null.

In his paper, "The Problem of Infinite Matter in Steady-State Cosmology," Schlegel found a clever way out. Instead of regarding the past as a completed aleph-null set of finite time intervals (to be sure, ideal instants in time form an aleph-one continuum, but Schlegel is concerned with those finite time intervals during which doublings of atoms occur), we can view both the past and the future as infinite in the inferior sense of "becoming" rather than completed. Whatever date is suggested for the origin of the universe (remember, we are dealing with the steady-state model, not with a "big-bang" or oscillating theory), we can always set an earlier date. In a sense there is a "beginning," but we can push it as far back as we please. There is also an "end," but we can push it as far forward as we please. As we go back in time, continually halving the number of atoms, we never halve them more than a finite number of times, with the result that their number never shrinks to less than aleph-null. As we go forward in time, doubling the number of atoms, we never double more than a finite number of times; therefore the set of atoms never grows larger than aleph-null. In either direction the leap is never made to a completed aleph-null set of time intervals. As a result the set of atoms never leaps to aleph-one and the disturbing contradiction does not arise.

Cantor was convinced that his endless hierarchy of alephs, each obtained by raising 2 to the power of the preceding aleph, represented all the alephs there are. There are none in between. Nor is there an Ultimate Aleph, such as certain Hegelian philosophers of the time identified with the Absolute. The endless hierarchy of infinities itself, Cantor argued, is a better symbol of the Absolute.

All his life Cantor tried to prove that there is no aleph between aleph-null and C, the power of the continuum, but he never found a proof. In 1938 Kurt Gödel showed that Cantor's conjecture, which became known as the "continuum hypothesis," could be assumed to be true, and that this could not conflict with the axioms of set theory.

What Cohen proved in 1963 was that the opposite could also be assumed. One can posit that C is *not* aleph-one; that there is at least one aleph between aleph-null and C, even though no one has the slightest notion of how to specify a set (for example, a certain subset of the transcendental numbers) that would have such a cardinal number. This too is consistent with set theory. Cantor's hypothesis is undecidable. Like the parallel postulate of Euclidean geometry, it is an independent axiom that can be affirmed or denied. Just as the two assumptions about Euclid's parallel axiom divided geometry into Euclidean and non-Euclidean, so the two assumptions about Cantor's hypothesis now divide the theory of infinite sets into Cantorian and non-Cantorian. It is even worse than that. The non-Cantorian side opens up the possibility of an infinity of systems of set theory, all as consistent as standard theory now is, and all differing with respect to assumptions about the power of the continuum.

Of course Cohen did no more than show that the continuum hypothesis was undecidable within standard set theory, even when the theory is strengthened by the axiom of choice. Many mathematicians hope and believe that some day a "self-evident" axiom, not equivalent to an affirmation or denial of the continuum hypothesis, will be found, and that when this axiom is added to set theory, the continuum hypothesis will be decided. (By "self-evident" they mean an axiom which all mathematicians will agree is "true.") Indeed, both Gödel and Cohen expect this to happen and are convinced that the continuum hypothesis is in fact false, in contrast to Cantor, who believed and hoped it was true. So far, however, these remain only pious Platonic hopes. What is undeniable is that set theory has been struck a gigantic cleaver blow, and exactly what will come of the pieces no one can say.

ADDENDUM

IN GIVING a binary version of Cantor's famous diagonal proof that the real numbers are uncountable, I deliberately avoided complicating it by considering the fact that every integral fraction between 0 and 1 can be represented as an infinite binary fraction in two ways. For example, ¼ is .01 followed by aleph-null zeroes, and also .001 followed by aleph-null ones. This raises the possibility that the list of real binary fractions might be ordered in such a way that complementing the diagonal would produce a number on the list. The constructed number would, of course, have a *pattern* not on the list, but could not this be a pattern which expressed, in a different way, an integral fraction on the list?

The answer is no. The proof assumes that all possible infinite binary patterns are listed, therefore every integral fraction appears *twice* on the list, once in each of its two binary forms. It follows that the constructed diagonal number cannot match either form of any integral fraction on the list.

In every base notation there are two ways to express an integral fraction by an aleph-null string of digits. Thus in decimal notation $1/4 = .2500000 \ldots = .2499999 \ldots$. Although it is not necessary for the validity of the diagonal proof in decimal notation, it is customary to avoid ambiguity by specifying that each integral fraction be listed only in the form that terminates with an endless sequence of nines, then the diagonal number is constructed by changing each digit on the diagonal to a different digit other than nine or zero.

Until I discussed Cantor's diagonal proof in *Scientific American*, I had not realized how strongly the opposition to this proof has persisted; not so much among mathematicians as among engineers and scientists. I received many letters attacking the proof. William Dilworth, an electrical engineer, sent me a clipping from the *LaGrange Citizen*, LaGrange, Ill., January 20, 1966, in which he is interviewed at some length about his rejection of Cantorian "numerology." Dilworth first delivered his attack on the diagonal proof at the International Conference on General Semantics, New York, 1963.

One of the most distinguished of modern scientists to reject Cantorian set theory was the physicist P. W. Bridgman. He published a paper about it in 1934, and in his *Reflections of a Physicist* (Philosophical Library, 1955) he devotes pages 99–104 to an uncompromising attack on transfinite numbers and the diagonal proof. "I personally cannot see an iota of appeal in this proof," he writes, "but it appears to me to be a perfect nonsequitur—my mind will not do the things that it is obviously expected to do if it is indeed a proof."

The heart of Bridgman's attack is a point of view widely held by philosophers of the pragmatic and operationalist schools. Infinite numbers, it is argued, do not "exist" apart from human behavior. Indeed, all numbers are merely names for something that a person *does*, rather than names of "things." Because one can count twenty apples, but cannot count an infinity of apples, "it does not make sense to speak of infinite numbers as 'existing' in the Platonic sense, and still less does it make sense to speak of infinite numbers of different orders of infinity, as does Cantor."

"An infinite number," declares Bridgman, "is a certain aspect of what one does when he embarks on carrying out a process . . . an infinite number is an aspect of a *program* of action."

The answer to this is that Cantor *did* specify precisely what one must "do" to define a transfinite number. The fact that one cannot carry out an infinite procedure no more diminishes the reality or usefulness of Cantor's alephs than the fact that one cannot fully compute the value of pi diminishes the reality or usefulness of pi. It is not, as Bridgman maintained, a question of whether one accepts or rejects the Platonic notion of numbers as "things." For an enlightened pragmatist, who wishes to ground all abstractions in human behavior, Cantorian set theory should be no less meaningful or potentially useful than any other precisely defined abstract system such as, say, group theory or a non-Euclidean geometry.

ANSWERS

WHICH OF the five ESP symbols cannot be drawn an aleph-one number of times on a sheet of paper, assuming ideal lines that do not overlap or intersect, and replicas that may vary in size but must be similar in the strict geometric sense?

Only the plus symbol cannot be aleph-one replicated. Figure 22 shows how each of the other four can be drawn an aleph-one number of times. In each case points on line segment *AB* form an aleph-one continuum. Clearly a set of nested or side-by-side figures can be drawn so that a different replica passes through each of these points, thus putting the continuum of points into one-to-one correspondence with a set of nonintersecting replicas. There is no comparable way to place replicas of the plus symbol so that they fit snugly against each other. The centers of any pair of crosses must be a finite distance apart (although this distance can be made as small as one pleases), forming a countable (aleph-null) set of points. The reader may enjoy devising a formal proof that aleph-one plus symbols cannot be drawn on a page. The problem is similar to one involving alphabet letters

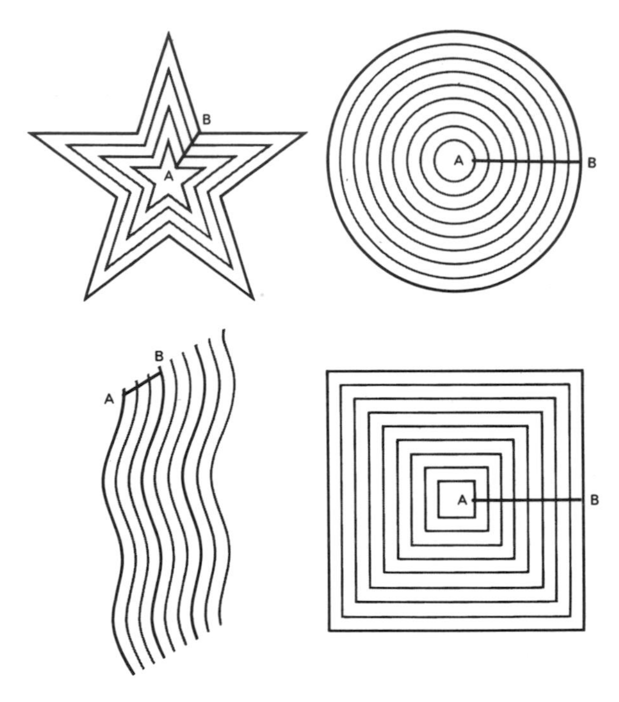

FIGURE 22
Proof for "ESP"-symbol problem

that can be found in Leo Zippin's *Uses of Infinity* (Random House, 1962), page 57. So far as I know, no one has yet specified precisely what conditions must be met for a linear figure to be aleph-one replicable. Some figures are aleph-one replicable by translation or rotation, some by shrinkage, some by translation plus shrinkage, some by rotation plus shrinkage. I rashly reported in my column that all figures topologically equivalent to a line segment or a simple closed curve were aleph-one replicable, but Robert Mack, then a high school student in Concord, Mass., found a simple counterexample. Consider two unit squares, joined like a vertical domino, then eliminate two unit segments so that the remaining segments form the numeral 5. It is not aleph-one replicable.

CHAPTER 4

Hypercubes

> *The children were vanishing.*
> *They went in fragments, like thick smoke in a wind, or like movement in a distorting mirror. Hand in hand they went, in a direction Paradine could not understand. . . .*
> —Lewis Padgett, from "Mimsy Were the Borogoves"

The direction that Paradine, a professor of philosophy, could not understand is a direction perpendicular to each of the three coordinates of space. It extends into four-space in the same way a chess piece extends upward into three-space with its axis at right angles to the x and y coordinates of the chessboard. In Padgett's great science fiction story, Paradine's children find a wire model of a tesseract (a hypercube of four dimensions) with colored beads that slide along the wires in curious ways. It is a toy abacus that had been dropped into our world by a four-space scientist tinkering with a time machine. The abacus teaches the children how to think four-dimensionally. With the aid of some cryptic advice in Lewis Carroll's *Jabberwocky* they finally walk out of three-space altogether.

Is it possible for the human brain to visualize four-dimensional structures? The 19th-century German physicist Hermann von Helmholtz argued that it is, provided the brain is given proper input data. Unfortunately our experience is confined to three-space and there is not the slightest scientific evidence that four-space actually exists. (Euclidean four-space must not be confused with the non-Euclidean four-dimensional space-time of relativity theory, in which time is handled as a fourth coordinate.) Nevertheless, it is conceivable that with the right kind of mathematical training a person might develop the ability to visualize a tesseract. "A man who devoted his life to it," wrote Henri Poincaré, "could perhaps succeed in picturing to himself a fourth dimension."

Charles Howard Hinton, an eccentric American mathematician who once taught at Princeton University and who wrote a popular book called *The Fourth Dimension*, devised a system of using colored blocks for making three-space models of sections of a tesseract. Hinton believed that by playing many years with this "toy" (it may have suggested the toy in Padgett's story), he had acquired a dim intuitive grasp of four-space. "I do not like to speak positively," he wrote, "for I might occasion a loss of time on the part of others, if, as may very well be, I am mistaken. But for my own part, I think there are indications of such an intuition. . . ."

Hinton's colored blocks are too complicated to explain here (the fullest account of them is in his 1910 book, *A New Era of Thought*). Perhaps, however, by examining some of the simpler properties of the tesseract we can take a few wobbly first steps toward the power of visualization Hinton believed he had begun to achieve.

Let us begin with a point and move it a distance of one unit in a straight line, as shown in Figure 23*a*. All the points on this unit line can be identified by numbering them from 0 at one end to 1 at the other. Now move the unit line a distance of one unit in a direction perpendicular to the line (*b*). This generates a unit square. Label one corner 0, then number the points from 0 to 1 along each of the two lines that meet at the zero corner. With these x and y coordinates we can now label every point on the square with an ordered *pair* of numbers. It is just as easy to visualize the next step. Shift the square a unit distance in a direction at right angles to both the x and the y axes (*c*). The result is a unit cube. With x, y, z coordinates along three edges that meet at a corner, we can label every point in the cube with an ordered *triplet* of numbers.

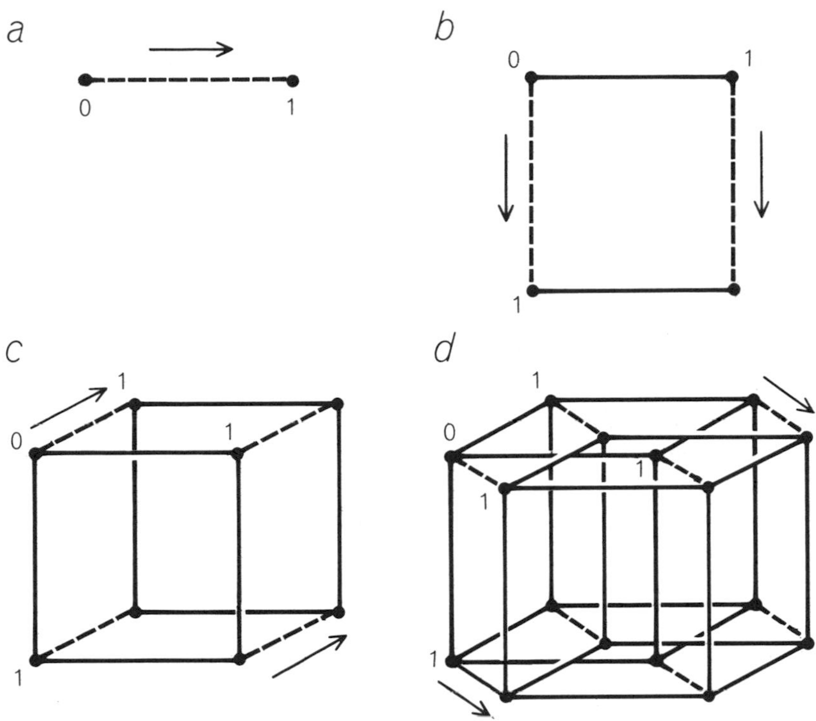

FIGURE 23
Steps toward generating a hypercube

Although our visual powers boggle at the next step, there is no logical reason why we cannot assume that the cube is shifted a unit distance in a direction perpendicular to all three of its axes (*d*). The space generated by such a shift is a four-space unit hypercube—a tesseract—with four mutually perpendicular edges meeting at every corner. By choosing a set of such edges as w, x, y, z axes, one might label every point in the hypercube with an ordered *quadruplet* of numbers. Analytic geometers can work with these ordered quadruplets in the same way they work with ordered pairs and triplets to solve problems in plane and solid geometry. In this fashion Euclidean geometry can be extended to higher spaces with dimensions represented by any positive integer. Each space is Euclidean but each is topologically distinct: a square cannot be continuously deformed to a straight line, a cube deformed to a square, a hypercube to a cube, and so on.

Accurate studies of figures in four-space can be made only on the basis of an axiomatic system for four-space, or by working analytically with the w, x, y, z equations of the four-coordinate system. But the tesseract is such a simple four-space structure that we can guess many of its properties by intuitive, analogical reasoning. A unit line has two end points. When it is moved to generate a square, its ends have starting and stopping positions and therefore the number of corners on the square is twice the number of points on the line, or four. The two moving points generate two lines, but the unit line has a start and a stop position and so we must add two more lines to obtain four as the number of lines bounding the square.

In similar fashion, when the square is moved to generate a cube, its four corners have start and stop positions and therefore we multiply four by two to arrive at eight corners on the cube. In moving, each of the four points generates a line, but to those four lines we must add the square's four lines at its start and the four lines at its stop, making $4 + 4 + 4 = 12$ edges on the cube. The four lines of the moving square generate four new faces, to which the start and stop faces are added, making $4 + 1 + 1 = 6$ faces on the cube's surface.

Now suppose the cube is pushed a unit distance in the direction of a fourth axis at right angles to the other three, a direction in which we cannot point because we are trapped in three-space. Again each corner of the cube has start and stop positions, so that the resulting tesseract has $2 \times 8 = 16$ corners. Each point generates a line, but to these eight lines we must add the start and stop positions of the cube's 12 edges to make $8 + 12 + 12 = 32$ unit lines on the hypercube. Each of the cube's 12 edges generates a square, but to those 12 squares we must add the cube's six squares before the push and the six after the push, making $12 + 6 + 6 = 24$ squares on the tesseract's hypersurface.

It is a mistake to suppose the tesseract is bounded by its 24 squares. They form only a skeleton of the hypercube, just as the edges of a cube form its skeleton. A cube is bounded by square faces and a hypercube by cubical faces. When the cube is pushed, each of its squares moves a unit distance in an unimaginable direction at right angles to its face, thereby generating another cube. To the six cubes generated by the six moving squares we must add the cube before it is pushed and the same cube after it is pushed, making eight in all. These eight cubes form the hypercube's hypersurface.

n-SPACE	POINTS	LINES	SQUARES	CUBES	TESSERACTS
0	1	0	0	0	0
1	2	1	0	0	0
2	4	4	1	0	0
3	8	12	6	1	0
4	16	32	24	8	1

FIGURE 24
Elements of structures analogous to the cube in various dimensions

The chart in Figure 24 gives the number of elements in "cubes" of spaces one through four. There is a simple, surprising trick by which this chart can be extended downward to higher n-cubes. Think of the nth line as an expansion of the binomial $(2x + 1)^n$. For example, the line segment of one-space has two points and one line. Write this as $2x + 1$ and multiply it by itself:

$$\begin{array}{r} 2x + 1 \\ 2x + 1 \\ \hline 4x^2 + 2x \quad\;\; \\ 2x + 1 \\ \hline 4x^2 + 4x + 1 \end{array}$$

Note that the coefficients of the answer correspond to the chart's third line. Indeed, each line of the chart, written as a polynomial and multiplied by $2x + 1$, gives the next line. What are the elements of a five-space cube? Write the tesseract's line as a fourth-power polynomial and multiply it by $2x + 1$:

$$\begin{array}{r}16x^4 + 32x^3 + 24x^2 + 8x + 1 \\ 2x + 1 \\ \hline 32x^5 + 64x^4 + 48x^3 + 16x^2 + 2x \\ 16x^4 + 32x^3 + 24x^2 + 8x + 1 \\ \hline 32x^5 + 80x^4 + 80x^3 + 40x^2 + 10x + 1\end{array}$$

The coefficients give the fifth line of the chart. The five-space cube has 32 points, 80 lines, 80 squares, 40 cubes, 10 tesseracts, and one five-space cube. Note that each number on the chart equals twice the number above it plus the number diagonally above and left.

If you hold a wire skeleton of a cube so that light casts its shadow on a plane, you can turn it to produce different shadow patterns. If the light comes from a point close to the cube and the cube is held a certain way, you obtain the projection shown in Figure 25. The network of this flat pattern has all the topological properties of the cube's skeleton. For example, a fly cannot walk along all the edges of a cube in a continuous path without going over an edge twice, nor can it do this on the projected flat network.

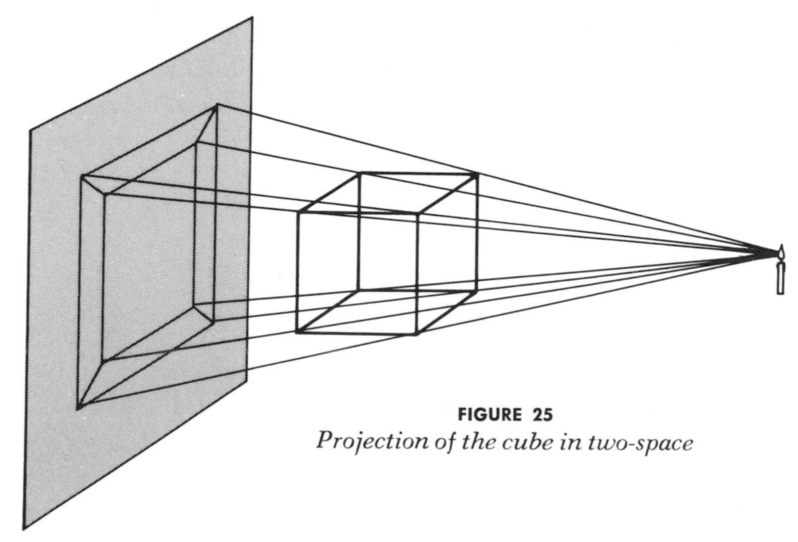

FIGURE 25
Projection of the cube in two-space

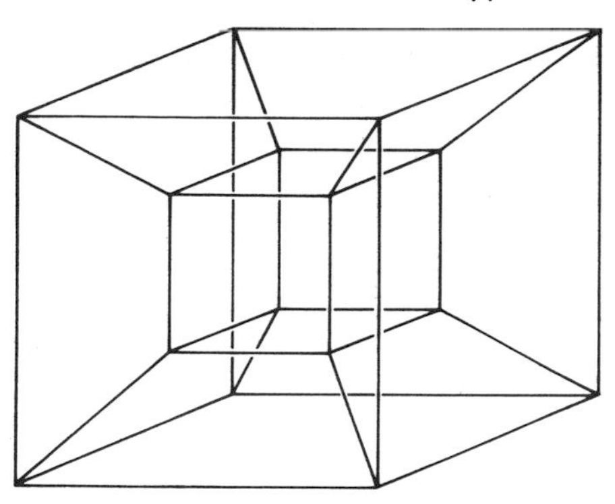

FIGURE 26
Projection of the tesseract in three-space

Figure 26 is the analogous projection in three-space of the edges of a tesseract; more accurately, it is a plane projection of a three-dimensional model that is in turn a projection of the hypercube. All the elements of the tesseract given by the chart are easily identified in the model, although six of the eight cubes suffer perspective distortions just as four of the cube's square faces are distorted in its projection on the plane. The eight cubes are the large cube, the small interior cube, and the six hexahedrons surrounding the small cube. (Readers should also try to find the eight cubes in Figure 23*d*—a projection of the tesseract, from a different angle, into another three-space model.) Here again the topological properties of both models are the same as those of the edges of the tesseract. In this case a fly *can* walk along all the edges without traversing any edge twice. (In general the fly can do this only on hypercubes in even spaces, because only in even spaces do an even number of edges meet at each vertex.)

Many properties of unit hypercubes can be expressed in simple formulas that apply to hypercubes of all dimensions. For example, the diagonal of a unit square has a length of $\sqrt{2}$. The longest diagonals on the unit cube have a length of $\sqrt{3}$. In general a diagonal from corner to opposite corner on a unit cube in n-space is $\sqrt{n}$.

A square of side x has an area of x^2 and a perimeter of $4x$. What size square has an area equal to its perimeter? The equation $x^2 = 4x$ gives x a value of 4. The unique answer is therefore a square of side 4. What size cube has a volume equal to its surface area? After the reader has answered this easy question he should have no difficulty answering two more: (1) What size hypercube has a hypervolume (measured by unit hypercubes) equal to the volume (measured by unit cubes) of its hypersurface? (2) What is the formula for the edge of an n-cube whose n-volume is equal to the $(n - 1)$-volume of its "surface"?

Puzzle books often ask questions about cubes that are easily asked about the tesseract but not so easily answered. Consider the longest line that will fit inside a unit square. It is obviously the diagonal, with a length of $\sqrt{2}$. What is the largest square that will fit inside a unit cube? If the reader succeeds in answering this rather tricky question, and if he learns his way around in four-space, he might try the more difficult problem of finding the largest cube that can be fitted into a unit tesseract.

An interesting combinatorial problem involving the tesseract is best approached, as usual, by first considering the analogous problems for the square and cube. Cut open one corner of a square [*see top drawing in Figure 27*] and its four lines can be unfolded as shown to form a one-dimensional figure. Each line rotates around a point until all are in the same one-space. To unfold a cube, think of it as formed of squares joined at their edges; cut seven edges and the squares can be unfolded (bottom drawing) until they all lie in two-space to form a hexomino (six unit squares joined at their edges). In this case each square rotates around an edge. By cutting different edges one can unfold the cube to make different hexomino shapes. Assuming that an asymmetric hexomino and its mirror image are the same, how many different hexominoes can be formed by unfolding a cube?

The eight cubes that form the exterior surface of the tesseract can be cut and unfolded in similar fashion. It is impossible to visualize how a four-space person might "see" (with three-dimensional retinas?) the hollow tesseract. Nevertheless, the eight cubes that bound it are true surfaces in the sense that the hyperperson can touch any point inside any cube with the point of a hyperpin without the pin's passing through any other point in the cube, just as we, with a pin, can touch any point inside

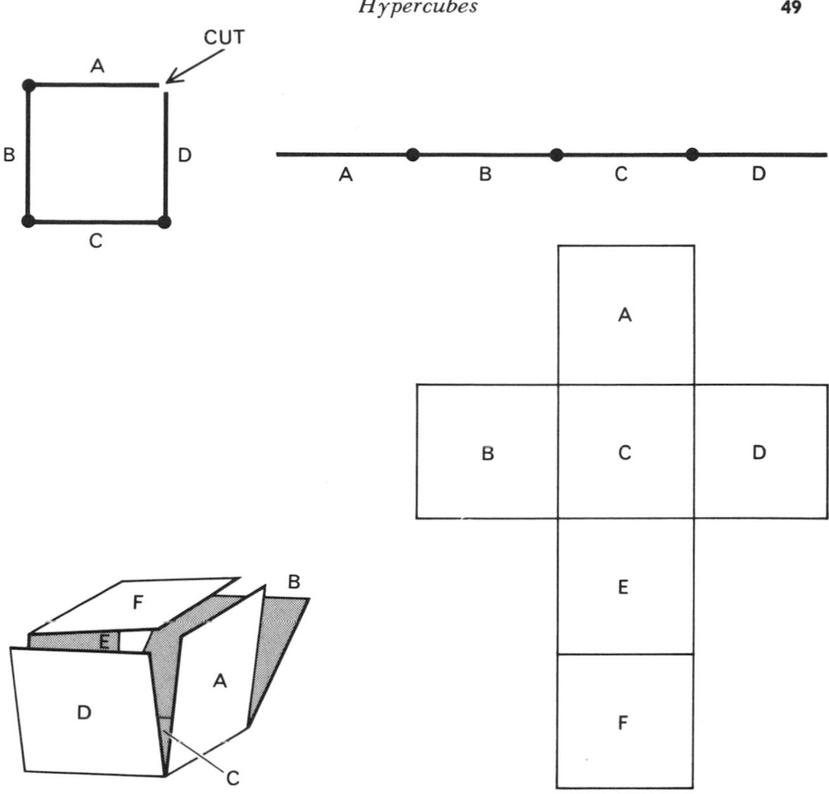

FIGURE 27
Unfolding a square (top) and a cube

any square face of a cube without the pin's going through any other point on that face. Points are "inside" a cube only to *us*. To a hyperperson every point in each cubical "face" of a tesseract is directly exposed to his vision as he turns the tesseract in his hyperfingers.

Even harder to imagine is the fact that a cube in four-space will rotate around any of its *faces*. The eight cubes that bound the tesseract are joined at their faces. Indeed, each of the 24 squares in the tesseract is a joining spot for two cubes, as can easily be verified by studying the three-space models. If 17 of these 24 squares are cut, separating the pair of cubes attached at that spot, and if these cuts are made at the right places, the eight cubes will be free to rotate around the seven uncut squares where they remain attached until all eight are in the same three-space. They will then form an order-8 polycube (eight cubes joined at their faces).

Salvador Dali's painting "Corpus Hypercubus" [*Figure 28, owned by the Metropolitan Museum of Art*] shows a hypercube unfolded to form a cross-shaped polycube analogous to the cross-shaped hexomino. Observe how Dali has emphasized the contrast between two-space and three-space by suspending his polycube above a checkerboard and by having a distant light cast shadows of Christ's arms. By making the cross an unfolded tesseract Dali symbolizes the orthodox Christian belief that the death of Christ was a metahistorical event, taking place in a region transcendent to our time and three-space and seen, so to speak, only in a crude, "unfolded" way by our limited vision. The use of Euclidean four-space as a symbol of the "wholly other" world has long been a favorite theme of occultists such as P. D. Ouspensky as well as of several leading Protestant theologians, notably the German theologian Karl Heim.

On a more mundane level the unfolded hypercube provides the gimmick for Robert A. Heinlein's wild story "—And He Built a Crooked House," which can be found in Clifton Fadiman's anthology *Fantasia Mathematica.* A California architect builds a house in the form of an unfolded hypercube, an upside-down version of Dali's polycube. When an earthquake jars the house, it folds itself up into a hollow tesseract. It appears as a single cube because it rests in our space on its cubical face just as a folded cardboard cube, standing on a plane, would appear to Flatlanders as a square. There are some remarkable adventures inside the tesseract and some unearthly views through its windows before the house, jarred by another earthquake, falls out of our space altogether.

The notion that part of our universe might fall out of three-space is not so crazy as it sounds. The eminent American physicist J. A. Wheeler has a perfectly respectable "dropout" theory to explain the enormous energies that emanate from quasi-

FIGURE 28

Salvador Dali's Corpus Hypercubus, *1954*

Metropolitan Museum of Art, gift of Chester Dale, 1955

stellar radio sources, or quasars. When a giant star undergoes gravitational collapse, perhaps a central mass is formed of such incredible density that it puckers space-time into a blister. If the curvature is great enough, the blister could pinch together at its neck and the mass fall out of space-time, releasing energy as it vanishes.

But back to hypercubes and one final question. How many different order-eight polycubes can be produced by unfolding a hollow hypercube into three-space?

ADDENDUM

HIRAM BARTON, a consulting engineer of Etchingham, Sussex, England, had the following grim comments to make about Hinton's colored cubes:

DEAR MR. GARDNER:

A shudder ran down my spine when I read your reference to Hinton's cubes. I nearly got hooked on them myself in the nineteen-twenties. Please believe me when I say that they are completely mind-destroying. The only person I ever met who had worked with them seriously was Francis Sedlak, a Czech neo-Hegelian philosopher (he wrote a book called The Creation of Heaven and Earth) *who lived in an Oneida-like community near Stroud, in Gloucestershire.*

As you must know, the technique consists essentially in the sequential visualizing of the adjoint internal faces of the polycolored unit cubes making up the large cube. It is not difficult to acquire considerable facility in this, but the process is one of autohypnosis and, after a while, the sequences begin to parade themselves through one's mind of their own accord. This is pleasurable, in a way, and it was not until I went to see Sedlak in 1929 that I realized the dangers of setting up an autonomous process in one's own brain. For the record, the way out is to establish consciously a countersystem differing from the first in that the core cube shows different colored faces, but withdrawal is slow and I wouldn't recommend anyone to play around with the cubes at all.

An attractive model of the hypercube, made of prepainted black and white aluminum strips, and designed to be hung as a mobile, was created and manufactured in 1972 by Eytan Kaufman, of New York City. Under the trade name Tesseract, it was sold by the Museum of Modern Art.

So far as I am aware, there has been no published solution to either of two problems which I conceived for my column, but for which I had no answer: (1) What is the largest cube that will fit inside a tesseract of unit side? (2) Into how many different order-8 polycubes can a hollow tesseract be cut and "unfolded" into three-space? I received several answers to the second question, and seven answers to the first. Unfortunately, no two solutions for either problem were in agreement, and I did not have the skill to evaluate any of them. Until an answer to either question is published and verified, both problems must be regarded as still unsolved.

ANSWERS

A TESSERACT of side x has a hypervolume of x^4. The volume of its hypersurface is $8x^3$. If the two magnitudes are equal, the equation gives x a value of 8. In general an n-space "cube" with an n-volume equal to the $(n - 1)$-volume of its "surface" is an n-cube of side $2n$.

The largest square that can be fitted inside a unit cube is the square shown in Figure 29. Each corner of the square is a distance of 1/4 from a corner of the cube. The square has an area of exactly 9/8 and a side that is three-fourths of the square root of 2. Readers familiar with the old problem of pushing the largest possible cube through a square hole in a smaller cube will recognize this square as the cross section of the limiting size of the square hole. In other words, a cube of side not quite three-fourths of the square root of 2 can be pushed through a square hole in a unit cube.

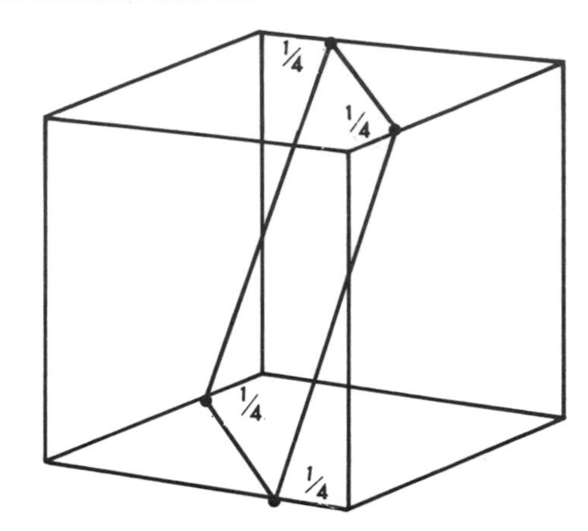

FIGURE 29
Packing a square in a cube

Figure 30 shows the 11 different hexominoes that fold into a cube. They form a frustrating set of the 35 distinct hexominoes, because they will not fit together to make any of the rectangles that contain 66 unit squares, but perhaps there are some interesting patterns they *will* form.

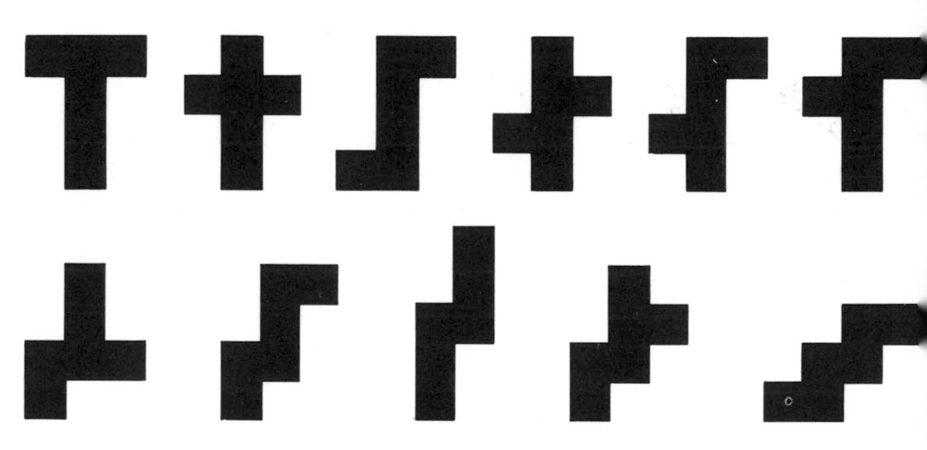

FIGURE 30
The 11 hexominoes that fold into cubes

CHAPTER 5

Magic Stars and Polyhedrons

"I've always been poor at geometry," he began. . . .
"You're telling me," said the demon gleefully.
Smiling flames, it came for him across the chalk lines of the useless hexagram Henry had drawn by mistake instead of the protecting pentagram.

—FREDERIC BROWN, from "Naturally," in *Honeymoon in Hell*

THE FIELD of combinatorial arithmetic has been getting increasing attention from mathematicians in recent decades, and along with this revival has come a new interest in combinatorial problems that were once considered mere puzzles. Herbert J. Ryser begins his excellent little book *Combinatorial Mathematics* (published in 1963 by the Mathematical Association of America) by displaying the three-by-three magic square, which was known in China centuries before the Christian era. "Many of the problems studied in the past for their amusement or aesthetic appeal are of great value today in pure and applied science," he writes. "Not long ago finite projective planes were regarded as a combinatorial curiosity. Today they are basic in the foundations of geometry and in the analysis and design of experiments. Our new technology with its vital concern with the discrete has given the recreational mathematics of the past a new seriousness of purpose."

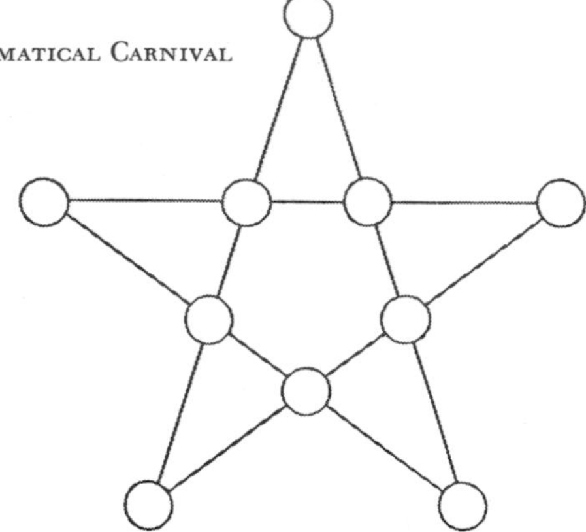

FIGURE 31
The pentagram

Magic squares are well known. In this chapter we will consider the less familiar but closely related topic of magic stars. It is a branch of recreational combinatorics that has a fascinating overlap with graph theory and the skeletal structure of polyhedrons.

The simplest star polygon is the familiar five-pointed Christmas star, which as children we learned to draw in one continuous path of five straight lines. It served as a recognition symbol for the ancient Greek Pythagoreans, and it was also their symbol of health. Old Greek coins often bore the symbol. In medieval and Renaissance witchcraft it was the mystic "pentagram" or "pentalpha." (The second name derives from the fact that it can be formed by superposing five capital A's.) The three large isosceles triangles that can also be superposed to make the star were taken as symbols of the Trinity and the star's points were often labeled J-E-S-U-S. When Goethe's Faust draws a pentagram on the threshold of his study, he fails to close the path. This slight break at one of the outer points allows Mephistopheles to enter the room, only to find himself trapped by the closed curve of the star's inner pentagon. Later, while Faust sleeps, the demon escapes by ordering a rat to nibble an opening in this pentagon.

Make a circle at each vertex of a pentagram [*see Figure 31*]. Is it possible to put the integers 1 through 10 in these 10 circles so that each line of four numbers will have the same sum? It is easy to determine what this sum, the "magic constant," must be. The numbers 1 through 10 sum to 55. Each number appears in *two* lines, therefore the sum of all five line sums must be twice 55, or 110. Since the five line sums are equal, each line must have a sum of 110/5, or 22. If a magic pentagram exists, therefore, its magic constant must be 22.

The fact that no such magic pentagram appears in the literature of witchcraft is strong evidence for its impossibility, and with a little ingenuity one can indeed prove that it cannot be accomplished. (See Harry Langman's *Play Mathematics*, 1962, pages 80–83.) The best we can do—without duplicating a number or using zero or negative numbers—is to label the vertices with 1, 2, 3, 4, 5, 6, 8, 9, 10, 12, as shown at the left in Figure 32. This makes a defective magic pentagram with the lowest possible constant, 24, and the lowest high number, 12.

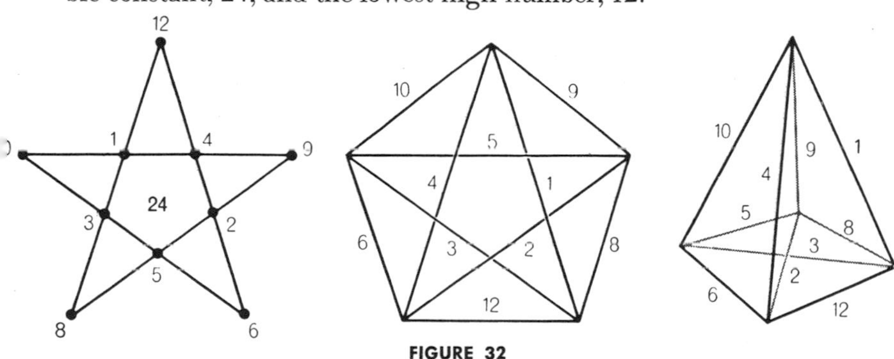

FIGURE 32
Pentagram (left) and equivalent graph (middle) and pentatope (right)

Consider now the following question, seemingly unrelated to the pentagram. Is it possible to label the 10 *edges* of the pentatope, or four-dimensional tetrahedron, so that at each corner the sum of the edges meeting at that corner is the same? Surprisingly, the question has just been answered; in its combinatorial aspect it is identical with the question about the pentagram! First draw the graph shown in the middle of Figure 32. This is called the "complete graph" for five points, because it joins each point to all the others. If you compare the numbers at the vertices of the pentagram with those on the edges of the graph, you will see that there is an identity of combinatorial structure. Every line of four numbers in the star matches a cluster of four

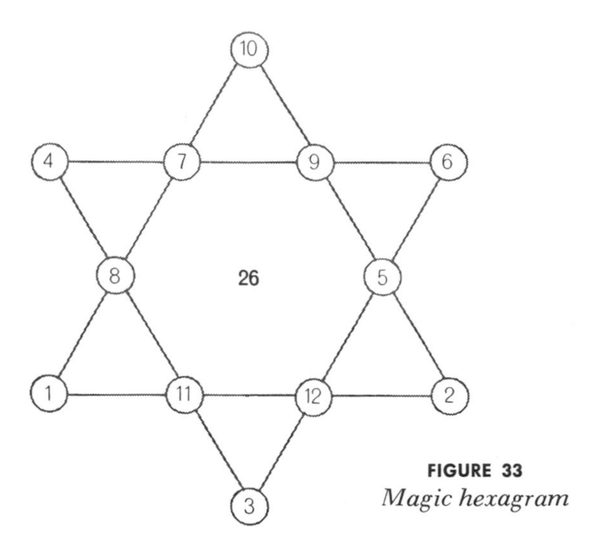

FIGURE 33
Magic hexagram

numbers on edges meeting at a common point. Because the magic star is impossible, it is impossible to make the complete graph for five points "magic" at its vertices.

Now, the complete graph for five points is topologically the same as the skeleton of the four-dimensional tetrahedron, as you can verify by comparing the numbers on the graph with those on a projection in three-space of the pentatope's skeleton [*at the right in Figure 32*]. A pentatope that is magic at the vertices is therefore not possible. Since the numbers shown on the pentatope's skeleton map back to those on the pentagram, we know that we have provided the pentatope with a nonconsecutive solution that has the lowest constant and lowest high number.

The situation grows in interest when we turn to the hexagram—also known as the hexalpha, Solomon's seal, and the Star of David [*see Figure 33*], a figure almost as prominent in the history of occultism and superstition as the pentagram. Because there are six lines, with each vertex common to two lines, and because the numbers 1 through 12 sum to 78, we obtain the magic constant $(2 \times 78)/6$, or 26. As the illustration shows, a magic hexagram is possible.

The problem of cataloguing all the different hexagram solutions, not counting rotations and reflections as being different, is not easy. One way to obtain new patterns is to transform the hexagram to its dual graph [*at left in Figure 34*], on which the same numbers mark lines that are magic at the vertices. It is easy to see that this graph is topologically the same as the skeleton of the octahedron (middle), one of the five Platonic solids. We can now rotate the octahedron and mirror-reflect it in any way we please, then map the numbers back to the hexagram (mapping edges to vertices according to the *original* numbering) and obtain new patterns for the hexagram.

Other transformations of the hexagram can be made, unrelated to rotations and reflections of the octahedron, that give still more solutions. Moreover, every magic star has what is called its "complement," obtained by replacing each number with the difference between that number and $n + 1$, when n is the highest of the star's consecutive integers. There are 80 different solutions, 12 with outer star points that also sum, as in the solution shown, to the constant.

There is still more to be said: The octahedron has what is called its "polyhedral dual," in which every face is replaced by a vertex and every vertex by a face, the edges remaining the same. The octahedron's dual is the cube. This allows us to label the 12 edges of a cube [*at the right in Figure 34*] with numbers 1 through 12 so that the cube is magic at its *faces;* that is, the sum of the four edges bounding every side is 26.

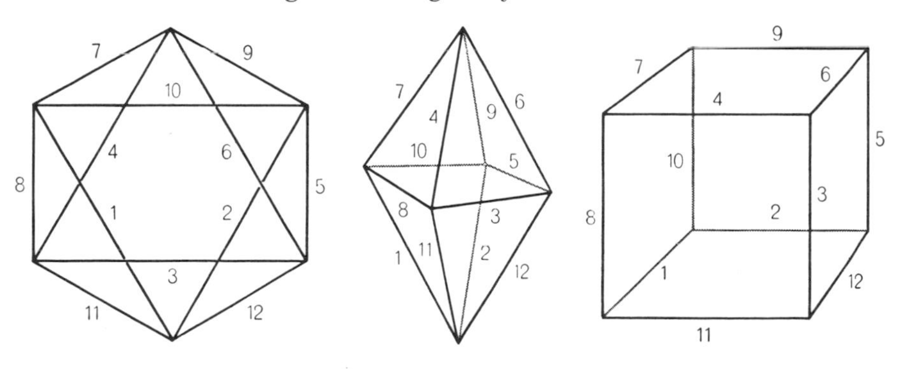

FIGURE 34
Graph of hexagram (left) and equivalent octahedron (middle) and cube (right)

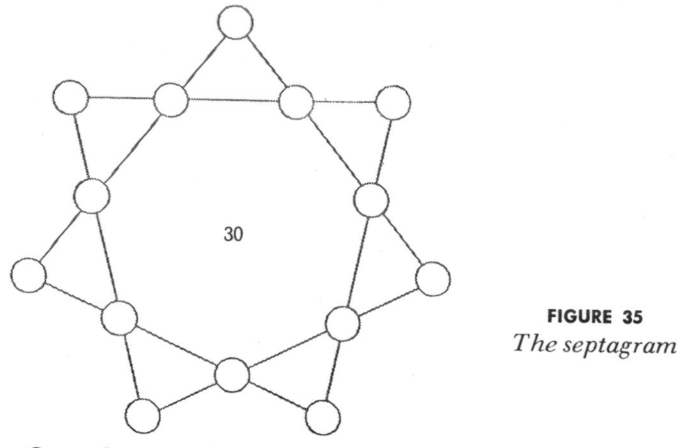

FIGURE 35
The septagram

Can the septagram, or seven-pointed star [*Figure 35*], be made magic by labeling its vertices with numbers 1 through 14? Yes, and I leave it to the reader to see how quickly he can find one of its 56 different solutions. The constant is $(2 \times 105)/7$, or 30. The best way to work on it is to draw a large figure, then put the numbers on small counters that can be slid over the paper. Warning: Once you start, you will find it hard to stop until you get a solution.

One solution for the octagram, or eight-pointed star, is shown at the left of Figure 36. Note that the magic constant, 34, is also the sum of the four corners of each of the two large squares. The top right drawing shows a corresponding graph that is magic at its vertices, and the drawing at the bottom right shows a solid with an equivalent skeleton. The octagram has 112 solutions.

Clearly there is no end to combinatorial problems that have to do with the labeling of edges, vertices, or faces of various polyhedrons so that magic constants are obtained in various ways. Many of these problems translate to equivalent magic-star problems. For example, which of the five regular solids can be made magic at their corners by being labeled along their edges with consecutive integers? It is easy to show that this is not possible on the tetrahedron. (See my *Sixth Book of Mathematical Games from Scientific American*, W. H. Freeman, 1971, page 194.) Is it possible on the cube? The cube's 12 edges [*Figure 37*], map to the 12 black spots on the vertices of the octagram (right). Since each spot is in two lines, the constant

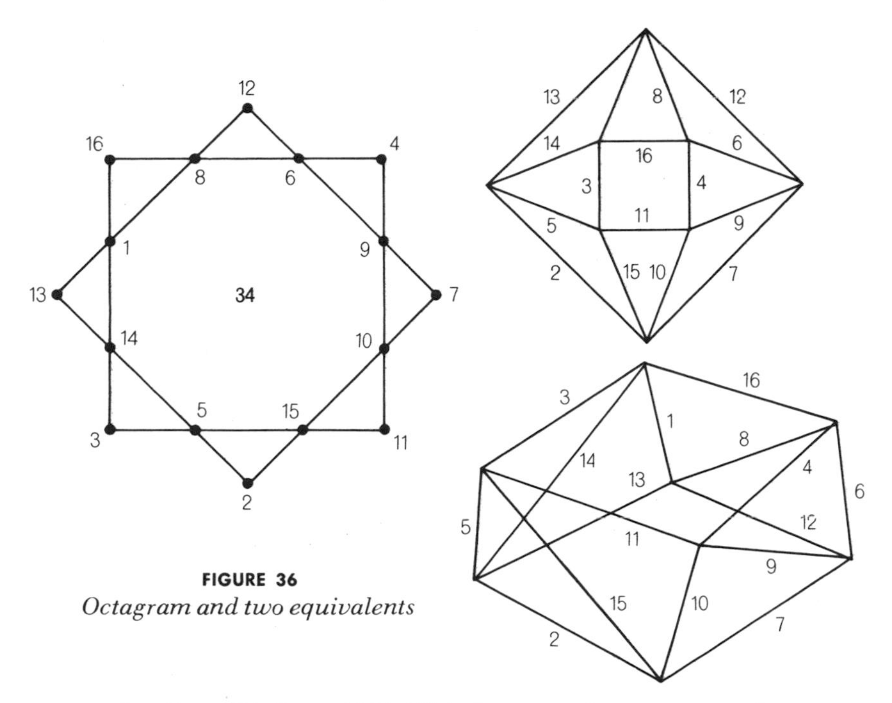

FIGURE 36
Octagram and two equivalents

must be $(2 \times 78)/8$, or $19\frac{1}{2}$. This is not an integer, so we know at once the problem has no solution. The best we can do is to label the spots (or the cube's edges) as shown, to get a defective solution with the lowest constant, 20, and the lowest high number. Since the octahedron is the polyhedral dual of the cube, this automatically solves the problem of labeling the octahedron's edges with different nonconsecutive, nonzero, positive integers to obtain the lowest constant at the faces.

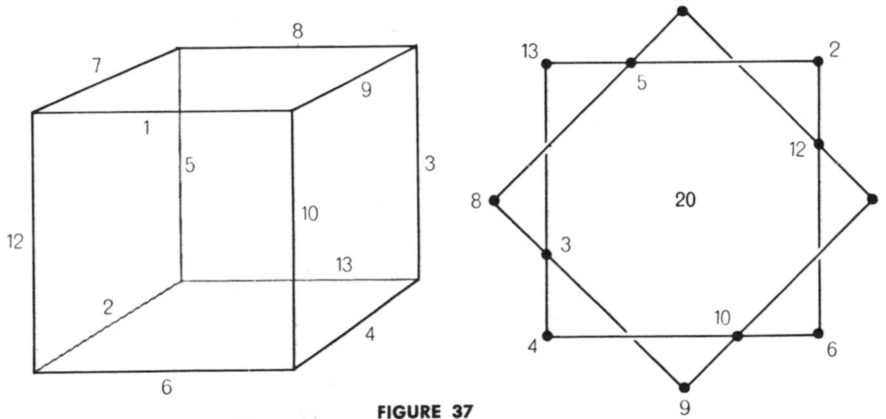

FIGURE 37
Magic cube skeleton (top) and octagram

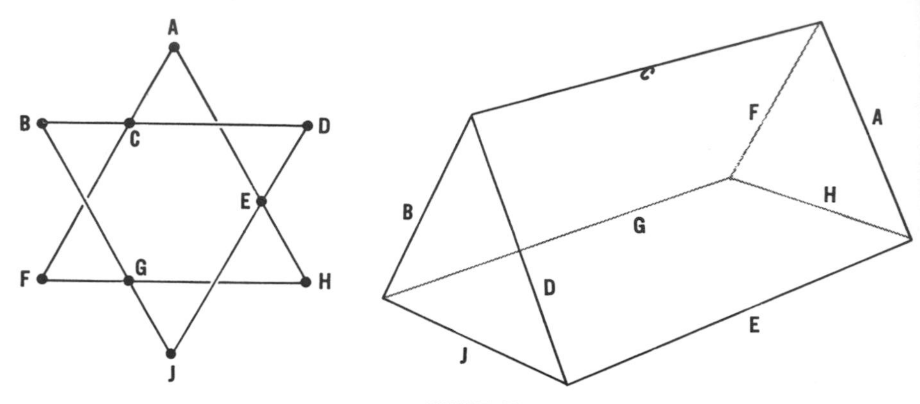

FIGURE 38
Is a magic triangular prism (right) possible? Equivalent star form is at left.

We have seen that the octahedron's edges can be labeled with consecutive integers to make it magic at its corners. On the icosahedron and dodecahedron the constant is not integral, so that they have no solutions. Since each is the other's polyhedral dual, there are no solutions for the corresponding problems of making them magic at their faces.

If we mark only nine of the hexagram's vertices as shown by the spots at the left of Figure 38, we obtain a magic-star problem equivalent to that of labeling the nine edges of a triangular prism (at right) with numbers 1 through 9 to make it magic at its six corners. The constant must be $(2 \times 45)/6$, or 15. Can it be done? It is not a difficult question.

ADDENDUM

Several readers sent crisp proofs of impossibility for the magic pentagram. Ian Richards, at the University of Minnesota, proved it this way:

1. Numbers 1 and 10 must be in the same line. Each of the two lines through 1 must contain three other numbers which add to 21, therefore the six numbers must add to 42. If 10 is not one of the six, the largest obtainable sum is $9 + 8 + 7 + 6 + 5 + 4 = 39$.

2. Let L be the line containing 1 and 10, L_1 the other line

through 1, and L_2 the other line through 10. L must contain one of four possible combinations. The quadruplet (1, 10, 4, 7) does not permit quadruplets for L_1 and L_2. The three possible combinations for L determine the quadruplets of the other two lines as follows:

L	L_1	L_2
1,10,2,9	1,6,7,8	10,5,4,3
1,10,3,8	1,5,7,9	10,6,4,2
1,10,5,6	1,4,8,9	10,7,3,2

3. Lines L_1 and L_2 must have one number in common. In each of the three possible cases, there is no such number. Therefore a magic five-pointed star is impossible.

The first attempt to enumerate the number of solutions for the six- and seven-pointed stars was made by H. E. Dudeney in *Modern Puzzles*, 1926, one of the two books reprinted in Dudeney's *536 Puzzles and Curious Problems* (Scribner's, 1967). He erred in both cases. E. J. Ulrich, of Enid, Okla., and A. Domergue, of Paris, found 80 patterns for the hexagram (six more than Dudeney). That the seven-pointed star has 72 solutions (as against Dudeney's 56) was first reported by Mrs. Peter W. Montgomery, of North Saint Paul, Minn. This was confirmed by Ulrich and Domergue, and later by a computer program written by Alan Moldon, at the University of Waterloo, Canada.

Domergue has reported 112 solutions for the eight-pointed star, and an estimate of more than 2,000 for the star with nine points. Results for stars of six, seven, and eight points were all confirmed in 1972 by Juan J. Roubicek, Buenos Aires. These figures exclude rotations and reflections, but include the complements.

ANSWERS

THE FIRST problem—to put the integers 1 through 14 on the vertices of a seven-pointed star so that every row of four numbers sums to 30—has 72 different solutions. One, with the first seven integers on the star's outer points, is shown in Figure 39.

The second problem was to determine if it is possible to label the nine edges of a triangular prism with the integers 1 through 9 so that the sum of the three edges meeting at every vertex is 15. The problem was shown to be the same as that of putting the nine digits on the spots of the hexagram in Figure 40 so that each row of three sums to 15.

Assume that there is a solution. Then:

$$A + C + F = A + E + H$$
$$C + F = E + H \qquad (1)$$

$$B + C + D = D + E + J$$
$$B + C = E + J \qquad (2)$$

$$F + G + H = J + G + B$$
$$F + H = J + B \qquad (3)$$

Combining (1) and (2), we write:

$$(C + F) - (B + C) = (E + H) - (E + J)$$
$$F - B = H - J \qquad (4)$$

Combining (3) and (4), we write:

$$(F - B) + (F + H) = (H - J) + (J + B)$$
$$2F - B + H = H + B$$
$$2F = 2B$$
$$F = B$$

But F cannot equal B because the problem called for nine *different* integers. One must therefore conclude that the original assumption is false and the problem has no solution. Note that this proves a much stronger result than the one asked for. It is

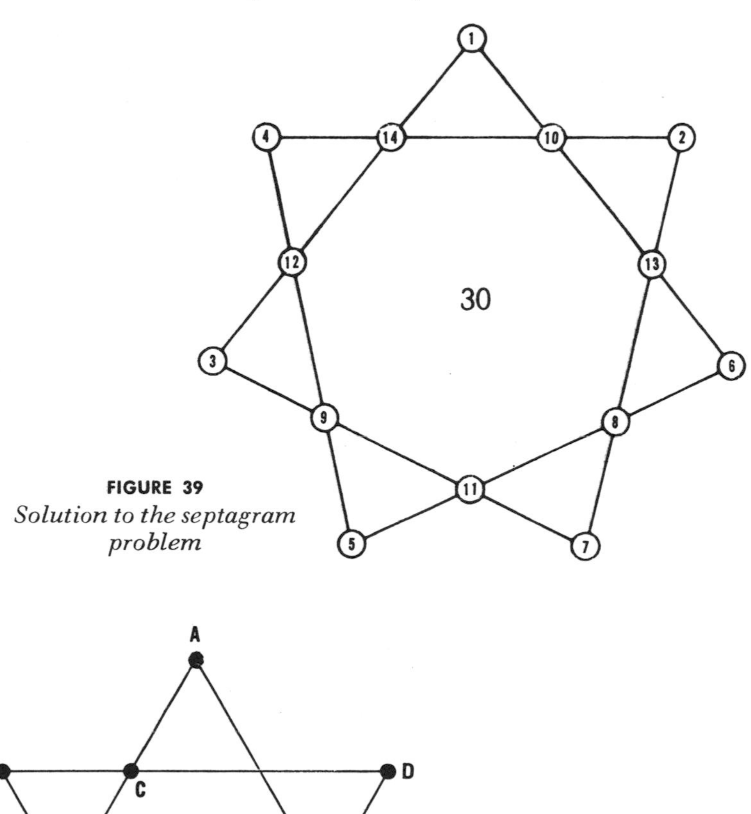

FIGURE 39
Solution to the septagram problem

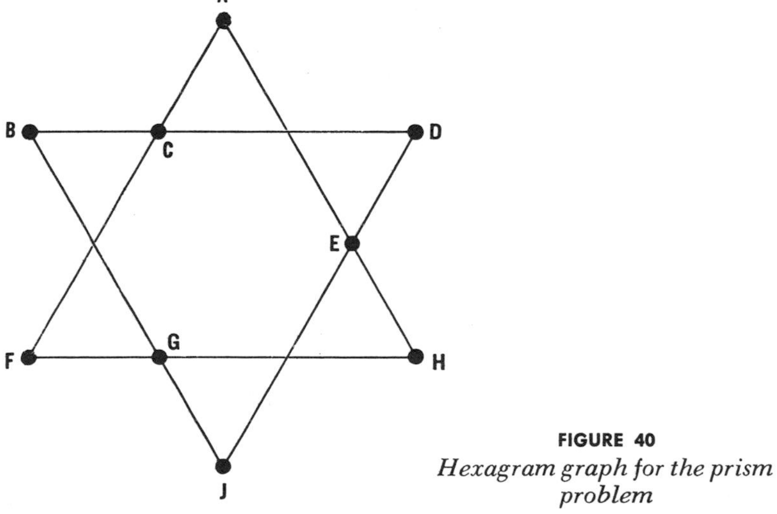

FIGURE 40
Hexagram graph for the prism problem

impossible to make the figure magic with different numbers of any kind whatever, consecutive or otherwise, rational or irrational.

CHAPTER 6

Calculating Prodigies

The ability to do arithmetic rapidly in one's head seems to have only a moderate correlation with general intelligence and even less with mathematical insight and creativity. Some of the most distinguished mathematicians have had trouble making change, and many professional "lightning calculators" (although not the best) have been dullards with respect to all other mental abilities.

Nevertheless, great mathematicians have also been skillful mental calculators. Carl Friedrich Gauss, for example, could perform prodigious feats of arithmetic in his mind. He liked to boast that he knew how to calculate before he could talk. When he was only three years old, his father, a bricklayer, was working on a weekly payroll for his laborers when young Friedrich startled him by saying, "Father, the reckoning is wrong. . . ." The boy gave a different sum, which proved to be correct when the long list of numbers was added again. No one had taught the child any arithmetic.

John von Neumann was a mathematical genius who was also gifted with this peculiar power to compute without pencil or paper. In his book *Brighter than a Thousand Suns* Robert Jungk tells of a meeting at Los Alamos during World War II at which ideas were tossed back and forth by von Neumann, Enrico Fermi, Edward Teller, and Richard Feynman. Whenever a mathematical calculation was called for, Fermi, Feynman, and von Neumann would spring into action. Fermi would do the work on a slide rule, Feynman would punch a desk calculator, and von Neumann would do it in his head. "The head," writes Jungk (quoting another physicist), "was usually first, and it is remarkable how close the three answers always checked."

The mental calculating abilities of Gauss, von Neumann, and other mathematical lions such as Leonhard Euler and John Wallis may seem miraculous; they pale, however, beside the feats of the professional stage calculators, a curious breed of mental acrobats who flourished throughout the 19th century in England, Europe, and America. Many began their careers as small boys. Although some wrote about their methods and were studied by psychologists, it seems likely that they held back most of their secrets or perhaps did not themselves fully understand how they did what they did.

The first of the stage calculators, Zerah Colburn, was born in 1804 in Cabot, Vt. Like his father, great-grandmother, and at least one brother, he had an extra finger on each hand and an extra toe on each foot. (The extra fingers were amputated when he was about 10. Did they stimulate, one wonders, his first efforts to count and calculate?) The child learned the multiplication table to 100 before he could read or write. His father, a poor farmer, was quick to see the commercial possibilities, and the lad was only six when his father took him on tour. His performances in England, when he was eight, are well documented. He could multiply any two four-digit numbers almost instantly, but he hesitated a moment on five-digit numbers. When told to multiply 21,734 by 543, he at once said 11,801,562. Asked how he had done it, he explained that 543 was equal to 181 times 3. Since it was easier to multiply by 181 than by 543, he had first multiplied 21,734 by 3, then multiplied the result by 181.

Washington Irving and other admirers of the boy raised enough money to send him to school, first in Paris and then in London. Either his calculating powers diminished thereafter or his interest in such feats declined. He returned to America when he was 20, and for about ten years was a Methodist circuit preacher. His quaint autobiography, *A Memoir of Zerah Colburn: written by himself . . . with his peculiar methods of calculation*, was published in Springfield, Mass., in 1833. At the time of his death at the age of 35 he was teaching foreign languages at Norwich University in Northfield, Vt. (He should not be confused with his nephew, of the same name, who wrote books on mechanical engineering, including a popular book, *The Locomotive Engine*.)

Colburn's stage career had its parallel in England in the performances of George Parker Bidder, born in 1806 in Devonshire. It is said that his father, a stonemason, taught him no more than how to count and that he acquired the ability to do arithmetic by playing with marbles and buttons. He was nine when he went on tour with his father. Typical of the kind of question put to him by strangers was: If the moon is 123,256 miles from the earth and sound travels four miles a minute, how long would it take for sound to travel (assuming that it could) from the earth to the moon? In less than a minute the boy replied: 21 days 9 hours 34 minutes. When asked (at age 10) for the square root of 119,550,669,121, he answered 345,761 in 30 seconds. In 1818, when he was 12 and Colburn was 14, the two boy wizards crossed paths in Derbyshire and were pitted against each other. Colburn implies in his memoirs that he won the contest, but London newspapers awarded the palm to Bidder.

Professors at the University of Edinburgh persuaded the elder Bidder to let them take over his son's education. The boy did well in college and eventually became one of England's most successful engineers. Most of his work had to do with railroads, but he is perhaps best known today as the man who designed and supervised the construction of the Victoria Docks in London. Bidder's calculating powers did not diminish with age. Shortly before his death in 1878 someone mentioned that there are 36,918 waves of red light per inch. Assuming that light travels at 190,000 miles per second, how many waves of red light, the man wondered, would strike the eye in one second. "You need not work it," Bidder said. "The number of vibrations will be 444,433,651,200,000."

Both Colburn and Bidder multiplied large numbers by breaking them into parts and multiplying from left to right by an algebraic crisscross technique often taught today in elementary schools that stress the "new math." For example, 236 × 47 is converted to (200 + 30 + 6) (40 + 7) and handled as shown in Figure 41. If the reader will close his eyes and try it, he will be surprised to find this method much easier to use in his head than the familiar right-to-left method. "True, the method . . . requires a much larger number of figures than the common Rule," Colburn wrote in his memoirs, "but it will be remembered that pen, ink and paper cost Zerah very little when engaged in a sum." (Throughout his book Colburn writes in the third person.) Why is this method so much easier to do in the head? Bidder, in a valuable lecture on his methods to the Insti-

PROBLEM: 236 × 47

236 = 200 + 30 + 6

47 = 40 + 7

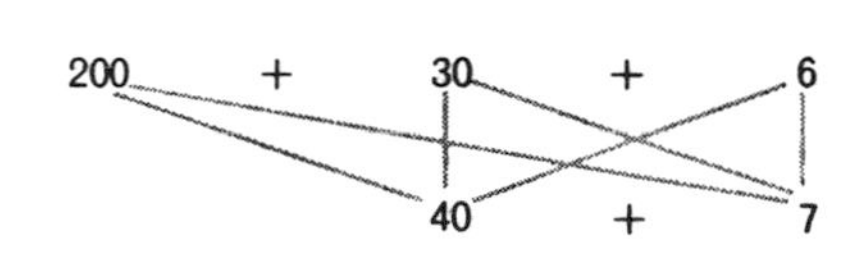

1. 40 × 200 = 8,000

2. 8,000 + (40 × 30) = 9,200

3. 9,200 + (40 × 6) = 9,440

4: 9,440 + (7 × 200) = 10,840

5. 10,840 + (7 × 30) = 11,050

6. 11,050 + (7 × 6) = 11,092

FIGURE 41

tute of Civil Engineers in London (published in 1856 in Volume 15 of the institute's *Proceedings*), gives the answer. After each step there is "one fact, and one fact only," that has to be held in the memory until the next step is completed.

Another reason why all stage calculators have preferred this method, although they seldom said so, is that they can start calling out a product while still calculating it. This is usually combined with other dodges to give the impression that computing time is much less than it really is. For example, a calculator will repeat a question, then answer it as though the result came into his mind immediately when actually he began calculating while the person was still calling out the second number. Sometimes he gains even more time by pretending not to hear the question so that it has to be repeated. One must bear these dodges in mind when reading any observer's account that speaks of a lightning calculator's "immediate" answers.

I shall pass quickly over the so-called idiot savants among the calculators. They were not so idiotic as their publicity made them out to be; besides, their speed was considerably less than that of stage performers with more intelligence. Jedediah Buxton, an 18th-century English farmer, was one of the earliest of the breed. He remained a farmer all his life and never gave public exhibitions, but local fame brought him to London to be tested by the Royal Society. Someone took him to the Drury Lane theater to see David Garrick in *Richard III*. Asked how he liked it, Buxton replied that the actors had spoken 14,445 words and taken 5,202 steps. Buxton had a compulsion to count and measure everything. He could walk over a field, it was said, and give an unusually accurate estimate of its area in square inches, which he would then reduce to square hairbreadths, assuming 48 hairs to an inch. He never learned to read, write, or work with written figures.

Perhaps the best all-around mental calculator of recent times was Alexander Craig Aitken, a professor of mathematics at the University of Edinburgh. He was born in New Zealand in 1895, and was coauthor of a classical textbook, *The Theory of Canoni-*

cal Matrices in 1932. Unlike most lightning calculators, he did not begin calculating mentally until he was 13, and then it was algebra, not arithmetic, that aroused his interest. In 1954, almost 100 years after Bidder's historic London lecture, Aitken spoke to the Society of Engineers in London on "The Art of Mental Calculation: With Demonstrations." His talk was published in the society's *Transactions* (December, 1954) to provide another valuable firsthand account of what goes on inside the mind of a rapid mental calculator.

A native ability to memorize numbers quickly is the one absolutely essential prerequisite. All the great stage calculators featured memory demonstrations. When Bidder was 10, he would ask someone to write a number of 40 digits and read it to him backward. He would at once repeat it forward. At the end of a performance many calculators could repeat accurately every number that had been involved. There are mnemonic tricks by which numbers can be transformed into words that in turn are memorized by other tricks (see my *Scientific American Book of Mathematical Puzzles and Diversions*, Chapter 11), but such techniques are much too slow for stage work and there is no question that the masters avoided such aids. "Mnemonics I have never used," Aitken said, "and deeply distrust. They merely perturb with alien and irrelevant association a faculty that should be pure and limpid."

Aitken mentioned in his lecture that he had recently read about how the contemporary French calculator Maurice Dagbert had been guilty of an "appalling waste of time and energy" when he had memorized pi to the 707 decimal places computed in 1873 by William Shanks. "It amused me to think," Aitken said, "that I had done this myself some years before Dagbert, and had found it no trouble whatever. All that had been necessary was to range the digits in rows of fifty each, each fifty being divided into ten groups of five, and to read these off in a particular rhythm. It would have been a reprehensibly useless feat had it not been so easy."

Twenty years later, after modern computers had carried pi

to thousands of decimal places, Aitken learned that poor Shanks had gone wrong on his last 180 digits. "I amused myself again," Aitken continued, "by learning the correct value as far as 1,000 places, and once again found it no trouble, except that I needed to 'fix' the join where Shanks's error had occurred. The secret, to my mind, is relaxation, the complete antithesis of concentration as usually understood. Interest is necessary. A random sequence of numbers, of no arithmetical or mathematical significance, would repel me. Were it necessary to memorize them, one might do so, but against the grain."

Aitken interrupted his lecture at this point by reciting pi, in an obviously rhythmic fashion, to 250 digits. Someone asked him to start the run at the 301st decimal. After he had given 50 digits he was asked to skip to the 551st place and give 150 more. He did all this without error, the digits being checked against a table of pi.

Do mental calculators visualize numbers while they work with them? Apparently some do and some don't, and some don't know whether they do or don't. The French psychologist Alfred Binet was on a committee of the Académie des Sciences that investigated the mental processes of two famous stage calculators of the late 19th century, a Greek named Pericles Diamandi and Jacques Inaudi, an Italian prodigy. In his 1894 book *Psychologie des grands calculateurs et joueurs d'échecs* Binet reported that Diamandi was a visualizer but that Inaudi, who was six times as fast, was of the auditory-rhythmic type. The visualizers have almost always been slower, although many professionals were of this type, such as Dagbert, the Polish calculator Salo Finkelstein, and a remarkable Frenchwoman who took the stage name of Mademoiselle Osaka. The auditory calculators such as Bidder seem to be more rapid. William Klein, a Dutch computer expert who used to perform under the name of Pascal (*Life* did a story about him in its issue of February 18, 1952), is probably the fastest living multiplier, capable of giving the product of two 10-digit numbers in less than two minutes. He too is an auditory calculator; indeed, he is unable to work without muttering rapidly to himself in Dutch. If he makes a mistake, it is usually caused by his confusing two numbers that *sound* alike. His brother Leo, almost as good a mental calculator, was a visual calculator who sometimes confused numbers that *look* alike.

Aitken said in his lecture that he could visualize if he wished; at various stages of calculation and at the finish the numbers sprang into visual focus. "But mostly it is as if they were hidden under some medium, though being moved about with decisive exactness in regard to order and ranging. I am aware in particular that redundant zeros, at the beginning or at the end of numbers, never occur intermediately. But I think that it is neither seeing nor hearing; it is a compound faculty of which I have nowhere seen an adequate description; though for that matter neither musical memorization nor musical composition in the mental sense have been adequately described either. I have noticed also at times that the mind has anticipated the will; I have had an answer before I even wished to do the calculation; I have checked it, and am always surprised that it is correct."

Aitken's skull housed an enormous memory bank of data. This is typical of the lightning calculators; I doubt that there has ever been one who did not know the multiplication table through 100, and some authorities have suspected that Bidder and others knew it to 1,000 but would not admit it. (Larger numbers can then be broken into pairs or triplets to be handled like single digits.) Long tables of squares, cubes, logarithms, and so on are stored in the memory along with countless numerical facts—such as the number of seconds in a year or ounces in a ton—that are useful in answering the kind of question audiences like to ask. Since 97 is the largest prime smaller than 100, calculators are often asked to compute the 96-digit recurring period for 1/97. Aitken long ago memorized it, so that if anyone popped *that* question he could rattle off the answer effortlessly.

There are in addition hundreds of shortcut procedures the calculator has learned to work out for himself. The first step in any complicated calculation, Aitken pointed out, is to decide in a flash on the best strategy. To illustrate, he disclosed a curious shortcut that is not well known. Suppose you were asked for the decimal reciprocal of a number ending in 9, say 59. Instead of dividing 1 by 59, you can add 1 to 59, making 60, then divide .1 by 6 in the manner shown in Figure 42. Note that at each step the digit obtained in the quotient is also entered in the dividend one place later. The result is the decimal for 1/59.

If asked to give the decimal for 5/23, Aitken went on, he realizes at once that he can multiply by 3 above and below the line to obtain the equivalent fraction 15/69, which has the desired 9 ending. He then changes 69 to 70, divides 1.5 by 7 according to the procedure just explained, and gets his answer. But he can also change the fraction to 65/299 and divide .65 by 3, entering the number *two* places further along in the dividend.

Which strategy is best? A decision has to be made instantly, Aitken said, and then followed by great steadfastness of purpose. Midway through the calculation it may flash into one's

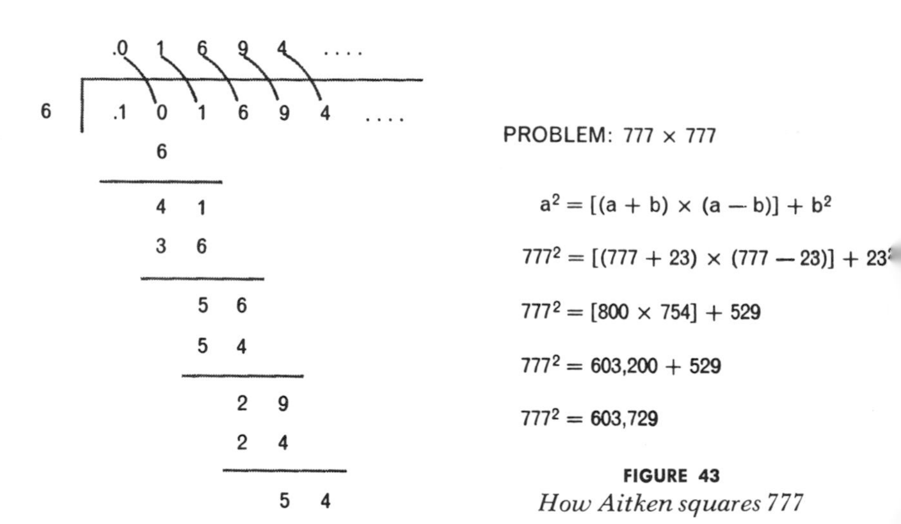

FIGURE 42
Aitken's computation of 1/59

FIGURE 43
How Aitken squares 777

mind that there is a better strategy. "One must resolutely ignore that, and keep on riding the inferior horse."

Aitken squared numbers by the method shown in Figure 43. The b is chosen to be fairly small and such that either $(a + b)$ or $(a - b)$ is a number ending in one or more zeros. In the case illustrated Aitken lets b equal 23. Having memorized a table of lower squares, he knows that 23^2 is 529 without thinking. During his lecture he was given seven three-digit numbers, each of which he squared almost instantly. Two four-digit numbers were squared in about five seconds. Note that Aitken's formula, when applied to any two-digit number ending in 5, leads to a delightfully simple rule that is worth remembering: Multiply the first digit by itself-plus-one and affix 25. For example, 85×85: 8 times 9 is 72, and appending 25 makes 7,225.

Thomas H. O'Beirne, a Glasgow mathematician, mentioned in a letter that he once went with Aitken to an exhibition of desk calculators. "The salesman-type demonstrator said something like 'We'll now multiply 23,586 by 71,283.' Aitken said right off 'And get . . .' (whatever it was). The salesman was too intent on selling even to notice, but his manager, who was watching, did. When he saw Aitken was right, he nearly threw a fit (and so did I)."

The machines are, of course, discouraging young people with wild talents like Aitken's from developing their skills. Aitken confessed at the close of his lecture that his own abilities began to deteriorate as soon as he acquired his first desk machine and saw how gratuitous his skill had become. "Mental calculators may, like the Tasmanian or the Moriori, be doomed to extinction," he concluded. "Therefore . . . you may be able to feel an almost anthropological interest in surveying a curious specimen, and some of my auditors here may be able to say in the year A.D. 2000, 'Yes, I knew one such.' "

In the next chapter I shall discuss some of the tricks of stage calculators by which even a tyro can obtain impressive results. Even the masters have not been above introducing pseudo-calculations into their stage work, much like an acrobat who gets applause for a showy feat that actually is not difficult at all.

ADDENDUM

SOLOMON W. GOLOMB often astounds his friends by evaluating in his head complicated expressions in combinatorial analysis. "The number of constants one need store in the memory," he writes, "and the number of simple rules, is far smaller than it seems." His greatest coup occurred when he was a college freshman. A biology teacher had just explained to the class that there are (as was then believed) 24 pairs of human chromosomes, therefore 2^{24} ways to select one member from each pair in the formation of an egg or sperm cell. "Thus from one parent," he said, "the number of different possible germ cells is 2^{24}, and you all know what that equals."

To this rhetorical question, Golomb immediately called out, "Yes, it's 16,777,216." The teacher laughed, looked down at his lecture notes, said, "Well, the *actual* value is . . . ," gasped, then demanded to know how Golomb had known it. Golomb replied that it was "obvious." The class immediately christened him "Einstein," and for the rest of the year several people, including the lab instructor, thought that was his name.

How *did* Golomb know it? He had (he tells me) recently memorized the values of n^n as far as $n = 10$. While the teacher was formulating his question, Golomb realized that $2^{24} = 8^8$, a number on his short list.

Contemporary lightning calculators do not make the headlines that they did in the nineteenth century, but a few are still in show business. Georgia-born Willis Nelson Dysart, who uses the stage name "Willie the Wizard," is best known in the United States. In Europe, the Indian lady calculator, Shakuntala Devi, and the French performer, Maurice Dagbert, are probably the most active, but my information is scanty concerning the stage calculators abroad.

CHAPTER 7

Tricks of Lightning Calculators

EVEN THE greatest of the lightning calculators discussed in the last chapter regularly included in their stage acts such feats as cube-root extraction and the calendar trick, which appear enormously difficult but actually are not, and many lesser calculators were not above introducing feats that operate almost entirely by trickery. Some of these tricks are so easily learned that the reader who wishes to amaze and confound his friends can master them with a minimum of practice and only the most elementary of calculating skills.

Consider, for example, the following multiplication trick, surprisingly little known even though it goes back to an Italian book of 1747, *I giochi numerici: fatti arcani* (*Numerical Games: Arcane Facts*), by G. A. Alberti. The trick works with numbers of any length, but it is best to limit it to three-digit numbers unless a pocket computer is handy for checking results.

Ask for any number with three digits. Suppose you are given 567. Write it twice on the blackboard or on a sheet of paper:

567　　　　567

Ask for another three-digit number. Write it under the 567 on the left. Now you need a different three-digit number as a multiplier on the right. It has to be (although your audience must not know this) the "9 complement" of the multiplier on the left; that is, corresponding digits of the two multipliers must add to 9. Assume that the left multiplier is 382. The right multiplier must be 617:

567	567
382	617

If you do the trick for a group, you can arrange beforehand for a friend to act as a secret confederate and suggest the correct second multiplier. Otherwise simply write it yourself as though you were putting down a number at random. Announce that you now intend to perform the two multiplications in your head, add the two products and then, as a final fillip, double the result. Obtain the sum of the two products instantly by subtracting 1 from the multiplicand and then appending the 9 complement. In this case 567 minus 1 is 566, the 9 complement of 566 is 433, and so the sum of the two products is 566,433. If you wrote this down, however, someone might notice that it began with the same two digits as the multiplicand, which would look suspicious, so you conceal the fact by doubling the number. This is not hard to do mentally, writing the digits from right to left as you perform the necessary doubling in your mind. If you prefer, you can mentally add a zero to 566,433 and divide by 5 (since multiplying by 10 and dividing by 5 is the same as multiplying by 2), in which case you write the final answer from left to right.

Why does the trick work? The sum of the two products is the same as the product of 567 and 999, which in turn is the same as multiplying 567 by 1,000 and subtracting 567. Do this on paper and you will see at once why the result has to be 566 followed by its 9 complement.

A subtler principle underlies a variety of lightning-multiplication tricks involving certain curious numbers that seem innocent enough but actually can be multiplied quickly by any number of equal or shorter length. Suppose the stage calculator asks for a nine-digit number and a confederate in the audience calls out 142,857,143. Another nine-digit number is requested and given legitimately. The performer multiplies the two numbers in his head, writing the mammoth product slowly from *left to right*. The secret is absurdly simple. Merely divide the second number through twice by 7. If there is a remainder after the first division, carry it back to the front of the first digit, then divide through a second time. Suppose the second number is 123,456,789. In effect, you divide 7 into 123,456,789,123,456,789. The result, 17,636,684,160,493,827, is the answer. The division must come out even; otherwise you know you have made a mistake.

The magic number 142,857,143 is just as easily multiplied by a number of shorter length. Merely add enough zeros at the end to make it a nine-digit number when you do the first mental division by 7. Thus if the multiplier were 123,456, you would, in your mind, divide 123,456,000,123,456 by 7. While you are writing the answer you are of course secretly looking at the multiplier and performing the mental division.

The number 142,857,143 was well known to the great stage calculators. One of the last of them to give vaudeville performances in the U.S. was the Indiana-born Arthur F. Griffith, who died in 1911 at the age of 31. He billed himself as "Marvelous Griffith" and had the reputation of being able to multiply two nine-digit numbers in less than half a minute. When I first read that, I became suspicious. Some digging at the library turned up an eyewitness account of his performance in 1904 before a group of students and faculty members at the University of Indiana. Griffith, the account says, wrote the number 142,857,143 on the blackboard. A professor was asked to put a nine-digit multiplier below it. As soon as he started to write it, from left to right, Marvelous Griffith began to write the product from left to right. "The student audience," the account continues, "rose with a shout." Griffith wrote a small book about his methods, *The Easy and Speedy Reckoner* (published in Goshen, Ind., in 1901), but it says nothing about 142,857,143.

There is one danger in using 142,857,143. If the multiplier happens to be evenly divisible by 7, the product "stutters"; that is, a series of numbers will be repeated in the answer and that will arouse suspicion. The performer may take a chance, knowing the odds are much in his favor, that the number won't stutter, and if it does, that the audience won't notice it. If he wants to avoid the stutter, he can mentally divide the multiplier by 7. If there is no remainder (hence a stutter) he can do any of several things. He can announce that, to make the feat even more incredible, he will reverse the multiplier, taking its digits in backward order, betting that the reversal is not a multiple of seven. Better still, he may ask the audience to further randomize the multiplier by altering one of its digits.

To avoid a stutter, Wallace Lee, a magician who invented many excellent mathematical tricks, devised the magic number 2,857,143. (It is the other number with its first two digits removed.) Ask for a seven-digit multiplier in which each digit is not less than 5. This, you explain, is to make the problem more difficult; actually it simplifies the procedure. The method is the same as before except that the entire multiplier must be doubled before you make the first division by 7. If all the digits are greater than 4, the doubling can be done in your head as you go along, digit by digit, in the following manner.

Assume that the multiplier is 8,965,797. Double the first digit, 8, and add 1, making 17. Seven goes into 17 twice, so you write 2 as the first digit of the answer, keeping the remainder, 3, in mind. Double the next number, 9, and add 1, making 19. Discard the first digit and substitute the 3 that was the previous remainder, making 39. Seven goes into 39 five times, so write 5 as the second digit of the answer, keeping the remainder, 4, in mind. Double the next number, 6, and add 1, making 13. Substitute 4 for the 1, making 43. Seven goes into 43 six times, so write 6 as the third digit of the answer and keep the remainder, 1, in mind. Double the next digit, 5, and add 1, making 11. Substituting 1 for 1 leaves the same number, 11, so you divide 11 by 7, getting 1 as the fourth digit of the answer and a remainder of 4 to keep in mind. Continue in this way until you reach the end of 8,965,797. When you double the last digit, 7, do *not* add 1.

The 2 that is the final remainder is carried back to the beginning to go in front of the 8. Now divide 8,965,797 by 7 in the ordinary manner, without doubling. The final result, 25,616,-564,137,971, is the desired product.

The doubling procedure used for the first division is not difficult to master. The product is guaranteed not to stutter and the trick's *modus operandi* is much harder for the uninitiated to discover. Like the previous magic number, this one can be multiplied by smaller numbers if you mentally add zeros to the multiplier. If the digits of the multiplier are not required to be greater than 4, a good procedure is to multiply the entire number by 10 (that is, add 0 to its end), then divide through by 35. This works because 35 is the product of 7 and half of 10. Of course you must memorize the multiples of 35.

Both tricks have such gargantuan products that unless an adequate desk calculator or pocket computer is available it is hard to get quick confirmation of your results. There are many smaller magic numbers, however, that work essentially the same way. For example, the product of 143 and *abc* is obtained by dividing *abc* through twice by 7 and hoping that the quotient does not stutter. The product of 1,667 and *abc* is obtained by adding a zero to *abc* and dividing through by 6, halving the remainder, if any (the remainder will be either 0, 2, or 4), carrying it back to the beginning and dividing *abc* by 3. This is very easy to do in your head, the result will not stutter, and spectators can check the answer without a machine—all of which makes it a capital impromptu trick to perform for friends.

The only reference I know to magic numbers of this type is in a privately printed work by the late Wallace Lee (he died in 1969) called *Math Miracles,* a book that contains many entertaining feats of lightning calculation. As a pleasant exercise in number theory the reader is asked to determine how the four magic numbers I have given were obtained and why they work the way they do.

In another impressive lightning-calculation feat you ask someone to cube any number from 1 to 100 and give you the result; you quickly name the cube root. To perform this trick it is necessary to memorize only the cubes of numbers 1 through 10 [*see Figure 44*]. Note that each cube ends in a different digit. (This is not true of squares, which explains why cube-root extraction is much easier for a calculator than square-root extraction.) The final digit matches the cube root in every case except 2, 3, 7, and 8. Those four exceptions are easily recalled because in each case the cube root and the final digit of the cube add to 10.

Suppose someone calls out the cube 658,503. Discard in your mind the last three digits and consider only what is left, 658. It lies between the cubes of 8 and 9. Pick the *lower* of the two, 8, and call out 8 as the first digit of the answer. The terminal digit of 658,503 is 3, so you know immediately that the second digit of the cube root is 7. Call out 7. The cube root is 87.

	CUBES	FIFTH POWERS
1	1	100 THOUSANDS
2	8	3 MILLIONS
3	27	24 MILLIONS
4	64	100 MILLIONS
5	125	300 MILLIONS
6	216	777 MILLIONS
7	343	1 BILLION, 500 MILLIONS
8	512	3 BILLIONS
9	729	6 BILLIONS
10	1,000	10 BILLIONS

FIGURE 44
Keys for root-extraction

Stage calculators often followed this trick by asking for fifth powers of numbers. This seems even harder than giving cube roots but in fact is both easier and faster. The reason is that the last digit of any fifth power of an integer always matches the last digit of the integer. Again, it is necessary to memorize a table [*see Figure 44*]. Suppose someone calls out 8,587,340,257. As soon as you hear "eight billion" you know that it lies between the ninth and the 10th number on the chart. Pick the lower number, 9. Ignore everything he says until he reaches the last digit, 7, at which point you instantly say 97. It is wise not to repeat this more than two or three times because it soon becomes obvious that final digits always match. The professional calculators worked with cubes and fifth powers of much larger numbers, by extensions of the systems given here, but I am limiting the explanation to the simpler two-digit roots.

The calendar trick—naming the day of the week for any date called out—was also featured by most of the great stage calculators. To perform it one must commit to memory the table shown in Figure 45, in which a digit is associated with each month. Initial memorization can be aided by the mnemonic cues on the right of the table, proposed by Wallace Lee in his book.

To calculate the day of the week in your head the following four-step procedure is recommended. There are other procedures, and even compact formulas, but this procedure is carefully designed for rapid mental computation.

1. Consider the last two digits of the year as a single number. Divide it mentally by 12 and keep the remainder in mind. You now add three small numbers: the number of dozens, the remainder, and the number of times 4 goes into the remainder. Example: 1910. Twelve goes into 10 no times, with a remainder of 10. Four goes into this remainder two times. $0 + 10 + 2 = 12$. If the final result is equal to or greater than 7, divide by 7 and remember only what is left. In the example given here, 12 divided by 7 has a remainder of 5, so only 5 is retained in the

JAN	1	THE **FIRST** MONTH.
FEB	4	A **C–O–L–D** (FOUR-LETTER) MONTH.
MAR	4	THE **K–I–T–E** MONTH.
APR	0	ON APRIL FOOLS' DAY I FOOLED **NO**BODY.
MAY	2	"MAY DAY" IS **TWO** WORDS.
JUN	5	THE **B–R–I–D–E** MONTH.
JUL	0	ON JULY 4 I SHOT **NO** FIRECRACKERS.
AUG	3	THE **H–O–T** MONTH.
SEP	6	START OF **A–U–T–U–M–N**.
OCT	1	A WITCH RIDES **ONE** BROOM.
NOV	4	A **C–O–O–L** MONTH.
DEC	6	BIRTH OF **C–H–R–I–S–T**.

FIGURE 45
Keys and mnemonic aids for calendar trick

mind. Henceforth this procedure will be called "casting out 7's." (A mathematician would say he was using "modulo 7" arithmetic.)

2. To the result of the preceding step add the month's key number. If possible, cast out 7's.

3. To the preceding result add the day of the month. Cast out 7's if possible. The resulting digit gives the day of the week, counting Saturday as 0, Sunday as 1, Monday as 2, and so on to Friday as 6.

4. If the year is a leap year and the month is January or February, go back one day from the final result.

The first step automatically alerts you to leap years. Leap years are multiples of 4 and any number is a multiple of 4 if its last two digits are. Therefore if there is no remainder when you divide by 12, or none when you make the division by 4, you know it is a leap year. (Bear in mind, however, that in the present Gregorian calendar system 1800 and 1900, although multiples of 4, are *not* leap years, whereas 2000 *is*. The reason is that the Gregorian calendar provides that a year ending in two zeros is a leap year only if it is evenly divisible by 400.)

The procedure just explained is restricted to dates in the 1900's, but only trivial final adjustments need to be made for dates in other centuries. For the 1800's go two days forward in the week. For the 2000's go one day back. It is best not to allow dates earlier than the 1800's because of confusion involving the shift that took place in England and the American colonies on September 14, 1752, from the Julian to the Gregorian calendar. Julius Caesar had used a year of 365.25 days, with a day added in February every fourth year to compensate for that excess fraction of one-fourth. Unfortunately the year has 365.2422+ days, so with the passage of centuries the leap years overcompensated and a sizable error of excess days accumulated. To prevent February from overtaking Easter (which depends on the vernal equinox), Pope Gregory XIII authorized the dropping of 10 days and the adoption of a calendar with fewer leap years. This was done throughout most of Europe in 1582, but in the English-speaking world the change was not made until 1752. The day after September 2 was called September 14, which explains why George Washington's birthday is now celebrated on February 22 instead of February 11, the actual date (Old Style) on which he was born. For dates in the 1700's, after the 1752 changeover, go forward four days in the week.

An example will make the procedure clear. Suppose you are informed that someone in the audience was born on July 28, 1929. What was the day of the week? Your mental calculations are as follows:

1. The 29 of 1929 contains two 12's, with a remainder of 5. Four goes once into 5. $2 + 5 + 1 = 8$. Casting out 7's reduces this to 1.

2. The key for July is 0, so nothing is added. The 1 is kept in mind.

3. The day of the month, 28, is added to 1. Cast out 7's from 29. The remainder is 1. Your subject was born on Sunday. (In actual practice this last step can be simplified by recognizing that 28 equals zero, modulo 7, so that there is nothing to add to the previous 1.)

The fourth step is omitted because 1929 is not a leap year. Even if it were, the step would still be left out because the month is not January or February, the only months for which leap year adjustments must be made.

From time to time so-called idiot savants get into the news by exhibiting an ability to perform this trick. A recent case of calendar-calculating twins with I.Q.'s in the 60-to-80 range was studied by psychiatrists and reported in *Scientific American* (August, 1965). It seems unlikely that any mysterious ability is operating in such cases. If the idiot savant takes a long time to give the day, he has probably memorized the first days of each year, over a wide range, and is simply counting forward in his mind from those key days to the given date. If he gives the day rapidly, he has probably been taught a method similar to the one I have described, or has come across it in a book or magazine.

Many methods for calculating the day of the week mentally were published late in the 19th century, but I have found none earlier than a method invented by Lewis Carroll and explained by him in *Nature* (Vol. 35, March 31, 1887, page 517). The method is essentially the same as the one described here. "I am not a rapid computer myself," Carroll wrote, "and as I find my average time for doing any such question is about 20 seconds, I have little doubt that a rapid computer would not need 15."

ADDENDUM

THE TRICK of doing rapid multiplication by mentally dividing has endless variants. One of Marvelous Griffith's favorite stunts was to multiply a large number by 125. Because $1/8 = .125$, you simply append three zeros and divide by 8.

The number 1,443 can be quickly multiplied by a two-digit number, *ab*, by dividing *ababab* by 7, and 3,367 can be multiplied by *ab* by dividing *ababab* by 3. (Reason: $1{,}443 = 10101/7$ and $3{,}367 = 10101/3$.)

The nonstuttering magic number 1,667 is my own discovery, as well as 8,335. To multiply 8,335 by a three-digit number, *abc*, append a zero to *abc* and divide by 12. Halve the remainder, if any, carry it back to the beginning, and (keeping zero at the end) divide by 6. It works because half of 1,667 is 833.5. I could think of no other four-digit numbers convenient to use in this manner. For a way of using the two numbers in a card trick, see my "Clairvoyant Multiplication" in the Indian magic periodical *Swami*, March, 1972, page 12.

Edgar A. Blair, Major W. H. Carter, and Kurt Eisemann each suggested a way of remembering the key numbers for the months that is probably easier for mathematicians than methods which employ key words. If grouped in triplets the keys are:

144 (Jan, Feb, Mar)
025 (Apr, May, Jun)
036 (Jul, Aug, Sep)
146 (Oct, Nov, Dec)

Note that the first three triplets are the squares of 12, 5, and 6, and that the last triplet, 146, is just 2 more than the first square.

ANSWERS

THE MAGIC numbers used in the lightning multiplication tricks operate on a principle best explained by examples. The number 142,857,143 is obtained by dividing 1,000,000,001 by 7. If 1,000,000,001 is multiplied by any nine-digit number, *abc,-def,ghi*, the product obviously will be *abc,def,ghi,abc,def,ghi*.

Therefore in order to multiply 142,857,143 by *abc,def,ghi* we have only to divide *abc,def,ghi,abc,def,ghi* by 7.

The second magic number, 2,857,143, is equal to 20,000,001 divided by 7. It is easy to see that in this case a seven-digit multiplier of 2,857,143 must be doubled before the first division by 7 is made. Insisting that each digit of the multiplier be greater than 4 (thus ensuring that there is always 1 to carry as each digit is doubled) makes possible the doubling procedure explained in the previous chapter. Without this proviso it is still possible to double and divide in the head, but the rules are more complicated

The smaller magic numbers 143 and 1,667 operate in similar ways. The first is equal to 1,001/7 and the second to 5,001/3. In the second case the multiplier, *abc*, must be multiplied by 5 before the first division by 3 is made. Since multiplying by 5 is the same as multiplying by 10 and dividing by 2, we add a zero to *abc* and divide by 6 as explained in the last chapter. The remainder must be halved, to convert it from sixths to thirds, and brought to the front for the second division, which is by 3. The fact that the second division is by a different number prevents the quotient from stuttering, something that always occurs if 143 is used and the multiplier, *abc*, happens to be a multiple of 7.

CHAPTER 8

The Art of M. C. Escher

What I give form to in daylight is only one percent of what I've seen in darkness.

—M. C. Escher

There is an obvious but superficial sense in which certain kinds of art can be called mathematical art. Op art, for instance, is "mathematical," but in a way that is certainly not new. Hard-edged, rhythmic, decorative patterns are as ancient as art itself, and even the modern movement toward abstraction in painting began with the geometric forms of the cubists. When the French Dadaist painter Hans Arp tossed colored paper squares in the air and glued them where they fell, he linked the rectangles of cubism to the globs of paint slung by the later "action" painters. In a broad sense even abstract expressionist art is mathematical, since randomness is a mathematical concept.

This, however, expands the term "mathematical art" until it becomes meaningless. There is another and more useful sense of the term that refers not to techniques and patterns but to a picture's subject matter. A representational artist who knows something about mathematics can build a composition around a mathematical theme in the same way that Renaissance painters did with religious themes or Russian painters do today with political themes. No living artist has been more successful with this type of "mathematical art" than Maurits C. Escher of the Netherlands.

"I often feel closer to mathematicians than to my fellow-artists," Escher has written, and he has been quoted as saying, "All my works are games. Serious games." His lithographs, woodcuts, wood engravings, and mezzotints can be found hanging on the walls of mathematicians and scientists in all parts of the world. There is an eerie, surrealist aspect to some of his work, but his pictures are less the dreamlike fantasies of a Salvador Dali or a René Magritte than they are subtle philosophical and mathematical observations intended to evoke what the poet Howard Nemerov, writing about Escher, called the "mystery, absurdity, and sometimes terror" of the world. Many of his pictures concern mathematical structures that have been discussed in books on recreational mathematics, but before we examine some of them, a word about Escher himself.

He was born in Leeuwarden in Holland in 1898, and as a young man he studied at the School of Architecture and Ornamental Design in Haarlem. For 10 years he lived in Rome. After leaving Italy in 1934 he spent two years in Switzerland and five in Brussels, then settled in the Dutch town of Baarn, where he and his wife now live. Although he had a successful exhibit in 1954 at the Whyte Gallery in Washington, he is still much better known in Europe than he is here. A large collection of his work is owned by Cornelius van Schaak Roosevelt of Washington, D.C., an engineer who is a grandson of President Theodore Roosevelt. It was through Roosevelt's generous cooperation, and with Escher's permission, that the pictures reproduced here were obtained.

Among crystallographers Escher is best known for his scores of ingenious tessellations of the plane. Designs in the Alhambra reveal how expert the Spanish Moors were in carving the plane into periodic repetitions of congruent shapes, but the Mohammedan religion forbade them to use the shapes of living things. By slicing the plane into jigsaw patterns of birds, fish, reptiles, mammals, and human figures, Escher has been able to incorporate many of his tessellations into a variety of startling pictures.

In *Reptiles*, the lithograph shown in Figure 46, a little monster crawls out of the hexagonal tiling to begin a brief cycle of three-space life that reaches its summit on the dodecahedron; then the reptile crawls back again into the lifeless plane. In *Day and Night*, the woodcut in Figure 47, the scenes at the left and the right are not only mirror images but also almost "negatives" of each other. As the eye moves up the center, rectangular fields flow into interlocking shapes of birds, the black birds flying into daylight, the white birds flying into night. In the circular woodcut *Heaven and Hell* [*Figure 48*] angels and devils fit together, the similar shapes becoming smaller farther from the center and finally fading into an infinity of figures, too tiny to be seen, on the rim. Good, Escher may be telling us, is a necessary background for evil, and vice versa. This remarkable tessellation is based on a well-known Euclidean model, devised by Henri Poincaré, of the non-Euclidean hyperbolic plane; the interested reader will find it explained in H. S. M. Coxeter's *Introduction to Geometry* (Wiley, 1961), pages 282–290.

If the reader thinks that patterns of this kind are easy to invent, let him try it! "While drawing I sometimes feel as if I were a spiritualist medium," Escher has said, "controlled by the creatures I am conjuring up. It is as if they themselves decide on the shape in which they choose to appear. . . . The border line between two adjacent shapes having a double function, the act of tracing such a line is a complicated business. On either side of it, simultaneously, a recognizability takes shape. But the human eye and mind cannot be busy with two things at the same moment and so there must be a quick and continuous jumping from one side to the other. But this difficulty is perhaps the very moving-spring of my perseverance."

It would take a book to discuss all the ways in which Escher's fantastic tessellations illustrate aspects of symmetry, group theory, and crystallographic laws. Indeed, such a book has been written by Caroline H. MacGillavry of the University of Am-

FIGURE 46
Reptiles, *lithograph, 1943*

FIGURE 47
Day and Night, *woodcut, 1938*

Mickelson Gallery, Washington

FIGURE 48
Heaven and Hell, *woodcut, 1960*

sterdam: *Symmetry Aspects of M. C. Escher's Periodic Drawings.* This book, published in Utrecht for the International Union of Crystallography, reproduces 41 of Escher's tessellations, many in full color.

Figures 49 and 50 illustrate another category of Escher's work, a play with the laws of perspective to produce what have been called "impossible figures." In the lithograph *Belvedere*, observe the sketch of the cube on a sheet lying on the checked floor. The small circles mark two spots where one edge crosses another. In the skeletal model held by the seated boy, however, the crossings occur in a way that is not realizable in three-space. The belvedere itself is made up of impossible structures. The youth near the top of the ladder is outside the belvedere but the base of the ladder is inside. Perhaps the man in the dungeon has lost his mind trying to make sense of the contradictory structures in his world.

FIGURE 49
Belvedere, *lithograph, 1958*

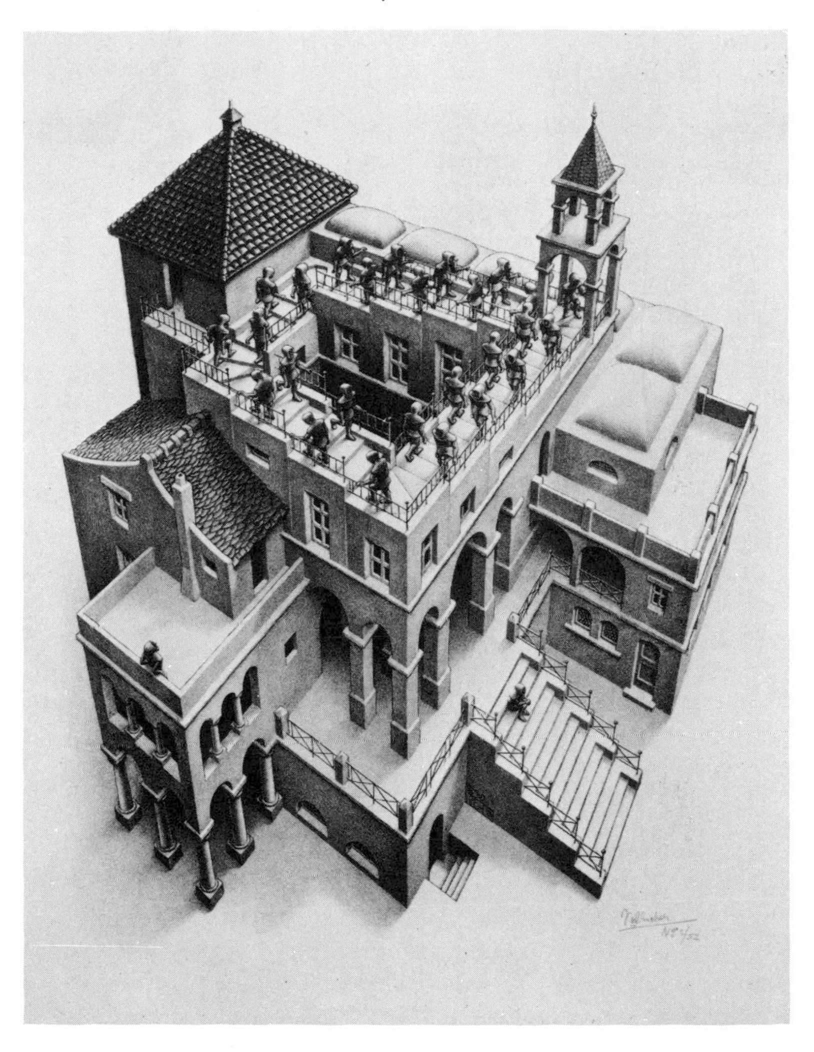

FIGURE 50
Ascending and Descending, *lithograph, 1960*

The lithograph *Ascending and Descending* derives from a perplexing impossible figure that first appeared in an article, "Impossible Objects: A Special Type of Visual Illusion," by L. S. Penrose, a British geneticist, and his son, the mathematician Roger Penrose (*British Journal of Psychology*, February, 1958). The monks of an unknown sect are engaged in a daily ritual of perpetually marching around the impossible stairway on the roof of their monastery, the outside monks climbing, the inside monks descending. "Both directions," comments Escher, "though not without meaning, are equally useless. Two refractory individuals refuse to take part in this 'spiritual exercise.' They think they know better than their comrades, but sooner or later they will admit the error of their nonconformity."

Many Escher pictures reflect an emotional response of wonder to the forms of regular and semiregular solids. "In the midst of our often chaotic society," Escher has written, "they symbolize in an unrivaled manner man's longing for harmony and order, but at the same time their perfection awes us with a sense of our own helplessness. Regular polyhedrons have an absolutely nonhuman character. They are not inventions of the human mind, for they existed as crystals in the earth's crust long before mankind appeared on the scene. And in regard to the spherical shape—is the universe not made up of spheres?"

The lithograph *Order and Chaos* [*Figure 51*] features the "small stellated dodecahedron," one of the four "Kepler-Poinsot polyhedrons" that, together with the five Platonic solids, make up the nine possible "regular polyhedrons." It was first discovered by Johannes Kepler, who called it "urchin" and drew a picture of it in his *Harmonices mundi* (*Harmony of the World*), a fantastic numerological work in which basic ratios found in music and the forms of regular polygons and polyhedrons are applied to astrology and cosmology. Like the Platonic solids, Kepler's urchin has faces that are equal regular polygons, and it has equal angles at its vertices, but its faces are not convex and they intersect one another. Imagine each of the 12 faces of the dodecahedron (as in the picture *Reptiles*) extended until it be-

FIGURE 51
Order and Chaos, *lithograph, 1950*

comes a pentagram, or five-pointed star. These 12 intersecting pentagrams form the small stellated dodecahedron. For centuries mathematicians refused to call the pentagram a "polygon" because its five edges intersect, and for similar reasons they refused to call a solid such as this a "polyhedron" because its faces intersect. It is amusing to learn that as late as the middle of the 19th century the Swiss mathematician Ludwig Schläfli, although he recognized some face-intersecting solids as being polyhedrons, refused to call this one a "genuine" polyhedron because its 12 faces, 12 vertices, and 30 edges did not conform to Leonhard Euler's famous polyhedral formula, $F + V = E + 2$. (It *does* conform if it is reinterpreted as a solid with 60 triangular faces, 32 vertices and 90 edges, but in this interpretation it cannot be called "regular" because its faces are isosceles triangles.) In *Order and Chaos* the beautiful symmetry of this solid, its points projecting through the surface of an enclosing bubble, is thrown into contrast with an assortment of what Escher has described as "useless, cast-off, and crumpled objects."

The small stellated dodecahedron is sometimes used as a shape for light fixtures. Has any manufacturer of Christmas tree ornaments, I wonder, ever sold it as a three-dimensional star to top a Christmas tree? A cardboard model is not difficult to make. H. M. Cundy and A. P. Rollett, in *Mathematical Models* (Oxford University Press, revised edition, 1961), advise one not to try to fold it from a net but to make a dodecahedron and then cement a five-sided pyramid to each face. Incidentally, every line segment on the skeleton of this solid is (as Kepler observed) in golden ratio to every line segment of next-larger length. The solid's polyhedral dual is the "great dodecahedron," formed by the intersection of 12 regular pentagons. For details about the Kepler-Poinsot star polyhedrons the reader is referred to the book by Cundy and Rollett and to Coxeter's *Regular Polytopes*.

The lithograph *Hand with Reflecting Globe* [*Figure 52*] exploits a reflecting property of a spherical mirror to dramatize what philosopher Ralph Barton Perry liked to call the "egocentric predicament." All any person can possibly know about the world is derived from what enters his skull through various sense organs; there is a sense in which one never experiences anything except what lies within the circle of his own sensations and ideas. Out of this "phenomenology" he constructs what he believes to be the external world, including those other people who appear to have minds in egocentric predicaments like his own. Strictly speaking, however, there is no way he can prove that anything exists except himself and his shifting sensations and thoughts. Escher is seen staring at his own reflection in the sphere. The glass mirrors his surroundings, compressing them inside one perfect circle. No matter how he moves or twists his head, the point midway between his eyes remains exactly at the

FIGURE 52
Hand with Reflecting Globe, *1935*

center of the circle. "He cannot get away from that central point," says Escher. "The ego remains immovably the focus of his world."

Escher's fascination with the playthings of topology is expressed in a number of his pictures. At the top of the woodcut *Knots* [*Figure 53*] we see the two mirror-image forms of the trefoil knot. The knot at top left is made with two long flat strips that intersect at right angles. This double strip was given a twist before being joined to itself. Is it a single one-sided band that runs twice around the knot, intersecting itself, or does it consist of two distinct but intersecting Möbius bands? The large knot below the smaller two has the structure of a four-sided tube that has been given a quarter-twist so that an ant walking inside, on one of the central paths, would make four complete circuits through the knot before it returned to its starting point.

The wood engraving *Three Spheres* [*Figure 54*], a copy of which is owned by New York's Museum of Modern Art, ap-

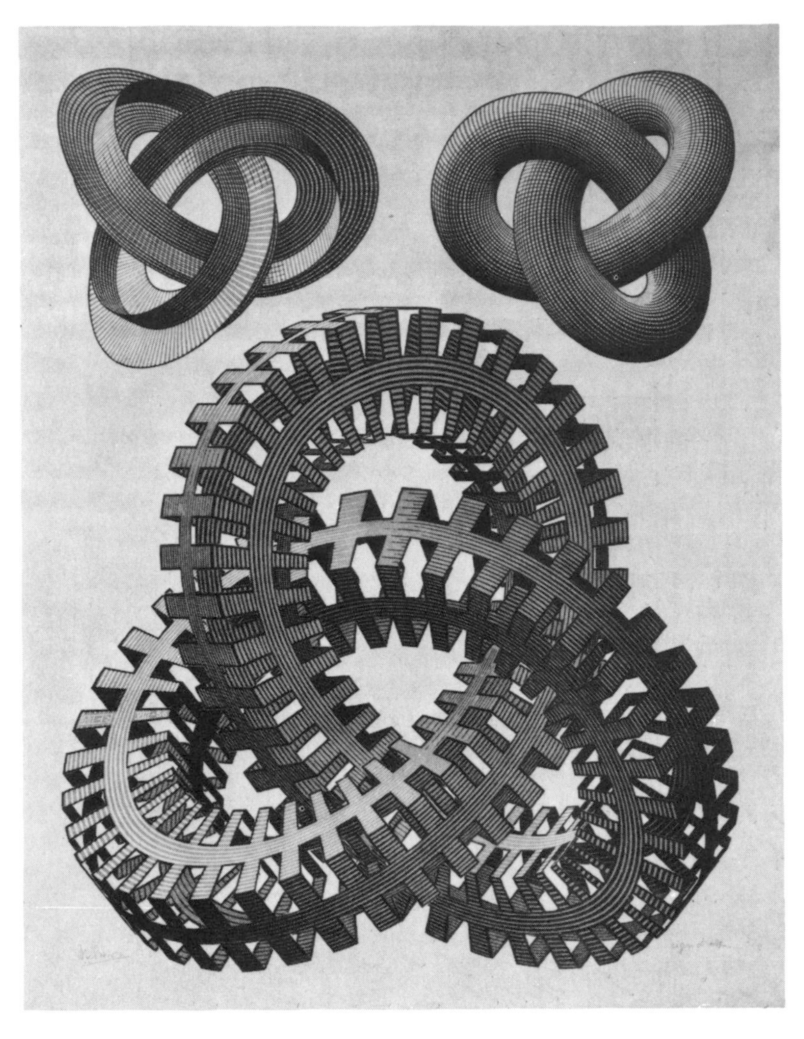

FIGURE 53
Knots, *woodcut, 1965*

pears at first to be a sphere undergoing progressive topological squashing. Look more carefully, however, and you will see that it is something quite different. Can the reader guess what Escher, with great verisimilitude, is depicting here?

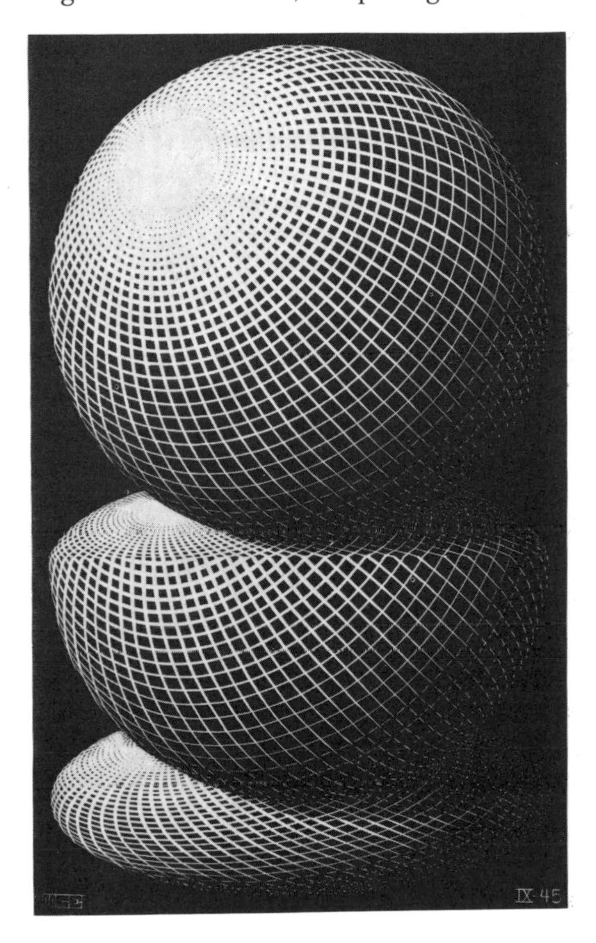

FIGURE 54
Three Spheres, *wood engraving, 1945*

ADDENDUM

When Escher died in 1972, at the age of 73, he was just beginning to become world-famous; not only among mathematicians and scientists (who were the first to appreciate him), but also with the public at large, especially with the young counterculture. Today the Escher cult is still growing. You see his pictures everywhere: on the covers of mathematical textbooks, on albums of rock music, on psychedelic posters that glow under black light, even on T-shirts. When I first reproduced an Escher picture in my column for April, 1961 (and *Scientific American* ran one of his bird tesselations on the cover), I purchased from Escher only one print, a woodcut. For a mere $40 to $60 each I could have bought scores of pictures that now would each be worth thousands. But who could then have anticipated the astonishing growth of Escher's fame?

So much has been written about Escher in recent years that I have made no attempt in my bibliography to list more than a few major books. The Abrams book contains the best and most complete reproductions of Escher's work, several essays on his art (including one by Escher himself), and an excellent list of selected references. Ken Wilkie's long article includes many previously unpublished Escher pictures as well as little-known details about the artist's private life and beliefs. *Holland Herald* is a newsmagazine in English, published in the Netherlands. (Subscription office: Medianet BV, Post Office Box 299, Haarlem, Netherlands.)

Cornelius Roosevelt's Escher collection is now owned by the National Gallery of Art, Washington, D.C.

ANSWERS

Three Spheres is a picture of three flat disks, each painted to simulate a sphere. The bottom disk is flat on a table. The middle disk is bent at right angles along a diameter. The top disk stands vertically on the horizontal half of the middle one. Clues are provided by a fold line in the middle disk and by identical shading on the three pseudospheres.

CHAPTER 9

The Red-Faced Cube and Other Problems

1. THE RED-FACED CUBE

RECREATIONAL MATHEMATICIANS have devoted much attention in the past to chessboard "tours" in which a chess piece is moved over the board to visit each square once and only once, in compliance with various constraints. John Harris, of Santa Barbara, has devised a fascinating new kind of tour—the "cube-rolling tour"—that opens up a wealth of possibilities.

To work on two of Harris' best problems, obtain a small wooden cube from a set of children's blocks or make one of cardboard. Its sides should be about the same size as the squares of your chessboard or checkerboard. Paint one side red. The cube is moved from one square to an adjacent one by being tipped over an edge, the edge resting on the line dividing the two cells. During each move, therefore, the cube makes one quarter-turn in a north, south, east, or west direction.

Problem 1. Place the cube on the northwest corner of the board, red side up. Tour the board, resting once only on every cell and ending with the cube red side up in the northeast corner. At no time *during* the tour, however, is the cube allowed to rest with the red side up. (NOTE: It is not possible to make such a tour from corner to diagonally opposite corner.)

Problem 2. Place the cube on any cell, an uncolored side up. Make a "reentrant tour" of the board (one that visits every cell once and returns the cube to its starting square) in such a way that at no time during the tour, including at the finish, will the cube's red side be up.

Both problems have unique solutions, not counting rotations and reflections of the path.

2. THE THREE CARDS

GERALD L. KAUFMAN, an architect and the author of several puzzle books, devised this logic problem.

Three playing cards, removed from an ordinary bridge deck, lie face down in a horizontal row. To the right of a King there's a Queen or two. To the left of a Queen there's a Queen or two. To the left of a Heart there's a Spade or two. To the right of a Spade there's a Spade or two. ("Two" means two cards, not a card with two spots.)

Name the three cards.

3. THE KEY AND THE KEYHOLE

THIS FRUSTRATING topological puzzle calls for a door key and a piece of heavy cord at least a few yards long. Double the cord, push the loop through the keyhole of a door as shown in the top drawing in Figure 55, then put both ends of the cord through the projecting loop as in the middle drawing. Now separate the ends, one to the left, the other to the right (bottom drawing). Thread the key on the left cord and slide it near the door, and secure the cord's ends by tying them to something—the backs of two chairs, for example. Allow plenty of slack.

The problem is to manipulate the key and cord so that the key is moved from spot *P* on the left to spot *Q* on the right. After the transfer the cord must be looped through the door in exactly the same way as before.

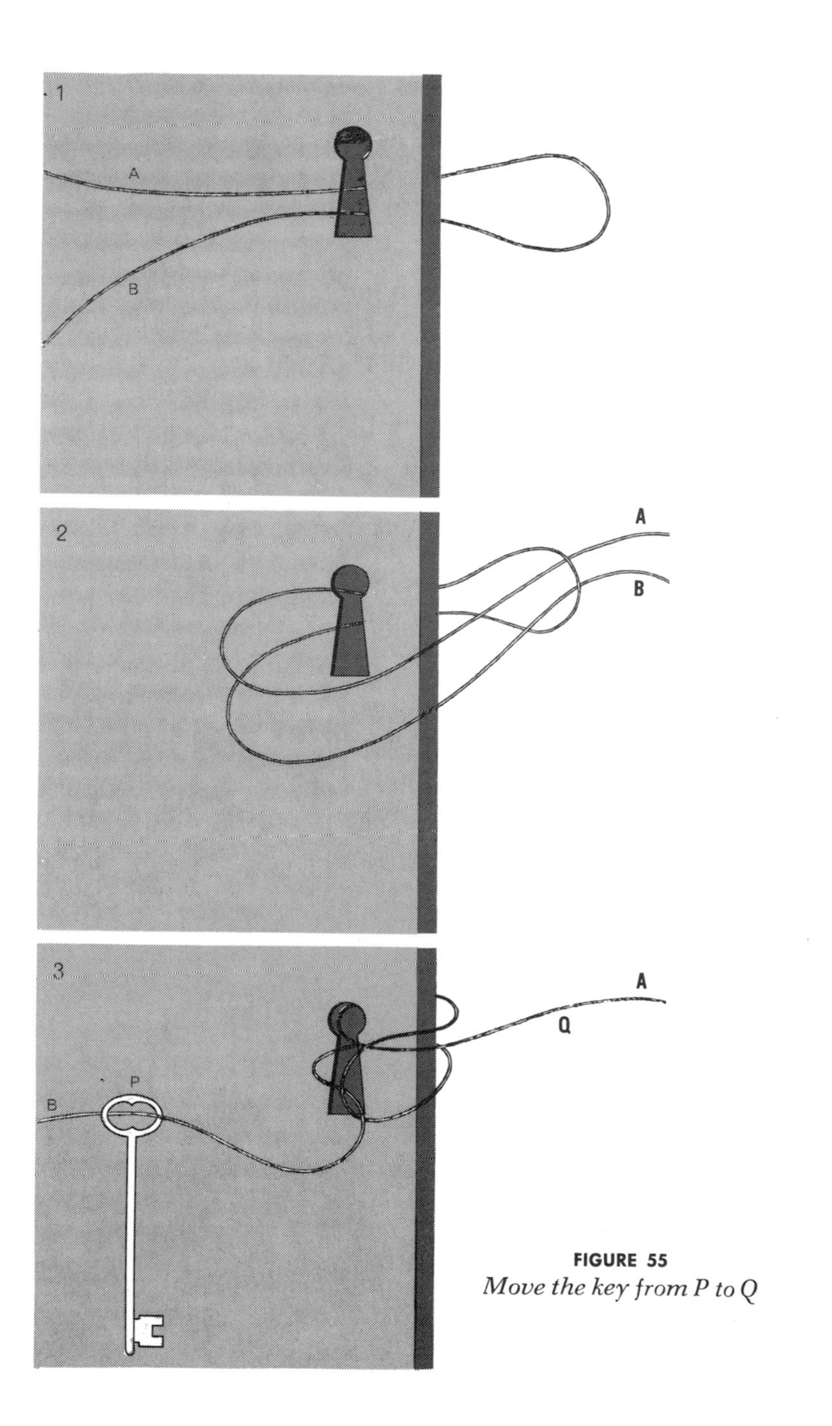

FIGURE 55
Move the key from P to Q

4. ANAGRAM DICTIONARY

NICHOLAS TEMPERLEY, while a student at Cambridge, proposed that devotees of wordplay produce, as a working tool, an anagram dictionary. Every word in English is first converted to its "alphabetical anagram," in which its letters appear in alphabetical order. SCIENTIFIC, for example, becomes CCEFIIINST. These alphabetical anagrams are then arranged in alphabetical sequence to form the dictionary. Each entry in the volume is an alphabetical anagram, and under it are listed all the English words that can be formed with those letters. Thus BDEMOOR will be followed by BEDROOM, BOREDOM, and all other words formed by those letters. AEIMNNOST will be followed by MINNESOTA and NOMINATES. Some entries even have mathematical interest. AEGILNRT, for instance, will be followed by such mathematical terms as INTEGRAL, RELATING, and TRIANGLE as well as other words such as ALTERING. EIINNSTXY will be followed by both NINETY-SIX and SIXTY-NINE. AGHILMORT will have among its anagrams both ALGORITHM and LOGARITHM. If a crossword puzzle gives the clue BEAN SOUP and an indication that this is an anagram, someone armed with the anagram dictionary need only alphabetize its letters and look up ABENOPSU to find SUBPOENA. If the clue is THE CLASSROOM, it takes only a moment to discover SCHOOLMASTER.

Most of the entries will begin with letters near the front of the alphabet. Temperley estimated that more than half will start with *A*, which is to say that more than half of all English words contain *A*. (This is not true of common words, but rarer words tend to be longer and are more likely to have an *A*.) After *I* the number of entries will drop sharply. Beyond *O* the list is extremely short.

Can the reader answer these questions?

1. What will be the dictionary's last entry? (Place names, such as Uz, the home of Job, are not included.)
2. What will be the first and second entries?

3. What will be the last entry starting with *A*?

4. What will be the first entry starting with *B*?

5. An entry, ABCDEFLO, begins with the first six letters of the alphabet. What is the word?

6. What will be the longest entry that is itself a word? (Short examples include ADDER, AGLOW, BEEFY, BEST, DIPS, FORT.)

7. What will be the longest entry that does not repeat any letter?

5. A MILLION POINTS

AN INFINITY of nontouching points lie inside the closed curve shown in Figure 56. Assume that a million of those points are selected at random. Will it always be possible to place a straight line on the plane so that it cuts across the curve, misses every point in the set of a million and divides the set exactly in half so that 500,000 points lie on each side of the line? The answer is yes; prove it.

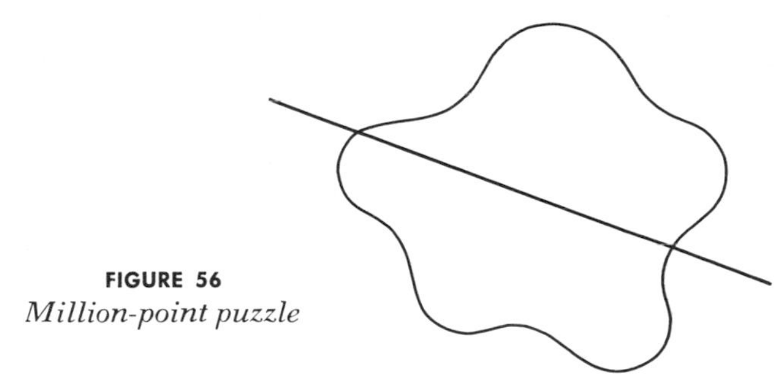

FIGURE 56
Million-point puzzle

6. LADY ON THE LAKE

A YOUNG lady was vacationing on Circle Lake, a large artificial body of water named for its precisely circular shape. To escape from a man who was pursuing her, she got into a rowboat and rowed to the center of the lake, where a raft was anchored. The man decided to wait it out on shore. He knew she would have to come ashore eventually. Since he could run four times as fast as she could row, he assumed that it would be a simple matter to catch her as soon as her boat touched the lake's edge.

But the girl—a mathematics major at Radcliffe—gave some thought to her predicament. She knew that once she was on solid ground she could outrun the man; it was only necessary to devise a rowing strategy that would get her to a point on shore before *he* could get there. She soon hit on a simple plan, and her applied mathematics applied successfully.

What was the girl's strategy? (For puzzle purposes it is assumed that she knows at all times her exact position on the lake.)

7. KILLING SQUARES AND RECTANGLES

THIS HARMLESS-LOOKING problem in combinatorial geometry, which I found on page 49 of *Sam Loyd and His Puzzles* (Barse and Co., 1928), has more to it than first meets the eye. Forty toothpicks are arranged as shown in Figure 57 to form the skeleton of an order-four checkerboard. The problem is to remove the smallest number of toothpicks that will break the perimeter of every square. "Every square" means not just the 16 small ones but also the nine order-two squares, the four order-three squares, and the one large order-four square that is the outside border—30 squares in all.

(On any square checkerboard with n^2 cells the total number of different rectangles is

$$\frac{(n^2+n)^2}{4},$$

of which

$$\frac{n(n+1)\,(2n+1)}{6}$$

are squares. "It is curious and interesting," wrote Henry Ernest Dudeney, the noted British puzzle expert, "that the total number of rectangles is always the square of the triangular number whose side is n.")

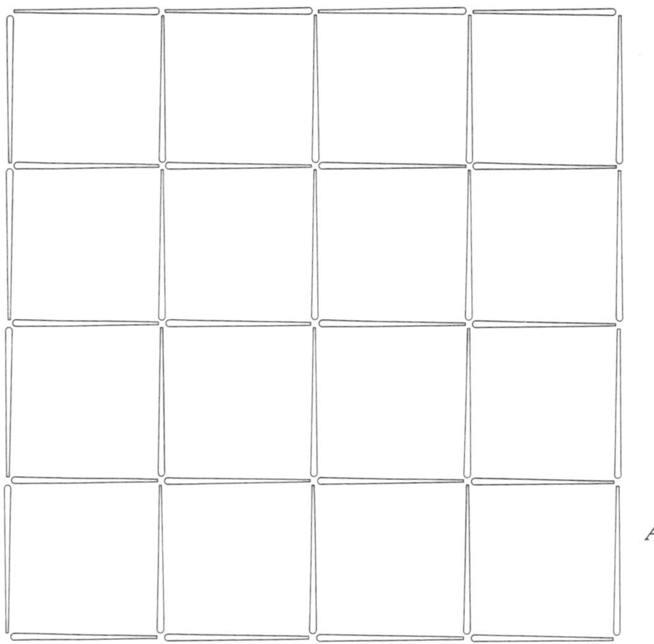

FIGURE 57
A toothpick puzzle

The answer given in the old book was correct, and the reader should have little difficulty finding it. But can he go a step further and state a simple proof that the answer is indeed minimum?

This far from exhausts the puzzle's depth. The obvious next step is to investigate square boards of other sizes. The order-one case is trivial. It is easy to show that three toothpicks must be removed from the order-two board to destroy all squares, and six from the order-three. The order-four situation is difficult enough to be interesting; beyond that the difficulty seems to increase rapidly.

The combinatorial mathematician is not likely to be content until he has a formula that gives the minimum number of toothpicks as a function of the board's order and also a method for producing at least one solution for any given order. The problem can then be extended to rectangular boards and to the removal of a minimum number of unit lines to kill all *rectangles*, including the squares. I know of no work that has been done on any of these questions.

The reader is invited to try his skill on squares with sides from four through eight. A minimal solution for order-eight, the standard checkerboard (it has 204 different squares), is not easy to find.

8. COCIRCULAR POINTS

FIVE PAPER rectangles (one with a corner torn off) and six paper disks have been tossed on a table. They fall as shown in Figure 58. Each corner of a rectangle and each spot where edges are seen to intersect marks a point. The problem is to find four sets of four "cocircular" points: four points that can be shown to lie on a circle.

For example, the corners of the isolated rectangle [*bottom right in Figure 58*] constitute such a set, because the corners of any rectangle obviously lie on a circle. What are the other three sets? This problem and the next are inventions of Stephen Barr, author of *Experiments in Topology* and *A Miscellany of Puzzles: Mathematical and Otherwise*, both published by Crowell.

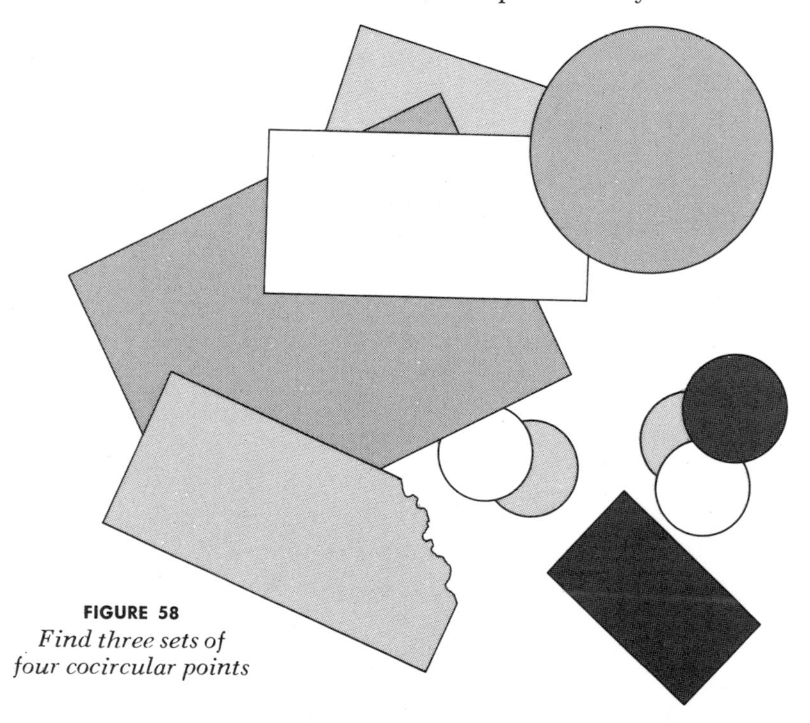

FIGURE 58
Find three sets of four cocircular points

9. THE POISONED GLASS

"Mathematicians are curious birds," the police commissioner said to his wife. "You see, we had all those partly filled glasses lined up in rows on a table in the hotel kitchen. Only one contained poison, and we wanted to know which one before searching that glass for fingerprints. Our laboratory could test the liquid in each glass, but the tests take time and money, so we wanted to make as few of them as possible. We phoned the university and they sent over a mathematics professor to help us. He counted the glasses, smiled and said:

" 'Pick any glass you want, Commissioner. We'll test it first.'

" 'But won't that waste a test?' I asked.

" 'No,' he said, 'it's part of the best procedure. We can test one glass first. It doesn't matter which one.' "

"How many glasses were there to start with?" the commissioner's wife asked.

"I don't remember. Somewhere between 100 and 200."

What was the exact number of glasses? (It is assumed that any group of glasses can be tested simultaneously by taking a small sample of liquid from each, mixing the samples and making a single test of the mixture.)

ANSWERS

1. Solutions to the cube-rolling tour problems are shown in Figure 59. In the first solution the red side of the cube is up only on the top corner squares. In the second, the dot marks the start of the tour, with the red side down.

Cube-rolling tours are a fascinating new field that, so far as I know, only John Harris has investigated in any depth. The problems one can devise are endless. Two of Harris' best: What reentrant tour has the red face on top as often as possible? Is there a reentrant tour that starts and ends with red on top but does not have red on top throughout the tour? One can invent problems in which more than one face is colored red, or faces may have different colors, or they may be marked with an *A* so

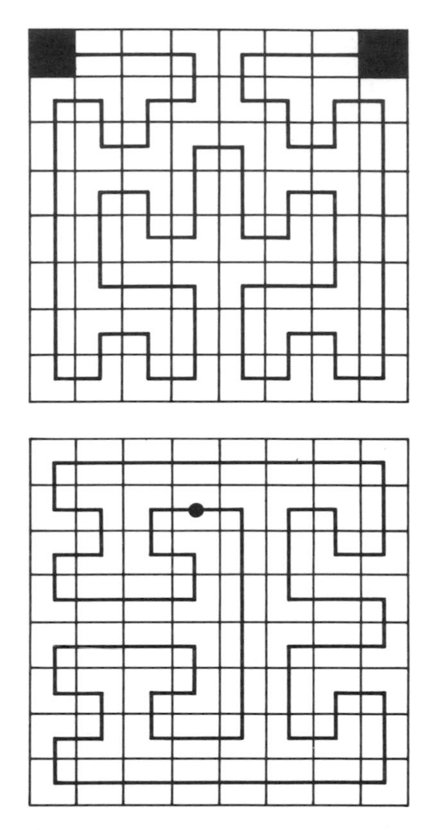

FIGURE 59
Cube-rolling solutions

that the orientation of the face is taken into account. What about rolling a standard die over a checkerboard to meet various provisos? Or a die numbered in a nonstandard way? See "Single Vacancy Rolling Cube Problems," by Harris, in *The Journal of Recreational Mathematics*, Vol. 7, Summer, 1974.

A novel board game based on rolling cubes was marketed in 1971 by Whitman (a subsidiary of Western Publishing Company) under the name of Relate. The board is a four-by-four checkerboard. The pieces are four identically colored cubes, two for each player. If each cube were numbered like a die, faces 1 and 2 would be one color, faces 3 and 5 a second color, 4 a third color, and 6 a fourth color. One player's pair of cubes is distinguished from the other player's by having a black spot on each face.

Play begins by alternately placing a cube on a cell, in any orientation, provided that each cube shows a different color on top. Assuming the cells are numbered left to right, one player puts his cubes on cells 3 and 4, the other on cells 13 and 14. These are called the player's starting cells. Players then take turns rolling one of their cubes to an orthogonally adjacent cell. There are three rules:

1. A player's two cubes must at all times have different colors on top.
2. If a player moves so the top color of his cube matches an opponent's cube, the opponent, on his next move, must move the matching cube to a new cell and so that it shows a new color on top.
3. If a move cannot be made without violating rules 1 and 2, a player must turn one of his cubes to a different color, but without leaving the cell it occupies. This counts as a move.

The winner is the first to occupy simultaneously the starting cells of his opponent. If one cube is on an opponent's starting cell, it still must move if forced by the opponent to do so.

I am indebted to John Gough, of Victoria, Australia, for calling this game to my attention. As he points out, the game suggests that there are unexplored possibilities for board games with rolling cubes on square lattices, or rolling tetrahedrons or octahedrons on triangular lattices.

2. The first two statements can be satisfied only by two arrangements of Kings and Queens: KQQ and QKQ. The last two statements are met by only two arrangements of Hearts and Spades: SSH and SHS. The two sets combine in four possible ways:

KS, QS, QH
KS, QH, QS
QS, KS, QH
QS, KH, QS

The last set is ruled out because it contains two Queens of Spades. Since each of the other three sets consists of the King of Spades, Queen of Spades and Queen of Hearts, we can be sure that those are the three cards on the table. We cannot know the position of any one card, but we can say that the first must be a Spade and the third a Queen.

3. To transfer the key from one side of the door to the other, first pass the key through the loop so that it hangs as shown at the left of Figure 60. Seize the double cord at points *A* and *B* and pull the loop back through and out of the keyhole. This will pull two new loops out of the hole, as shown in the middle. Move the key up along the cord, through both projecting loops. Grasp the two cords on the opposite side of the door and pull the two loops back through the hole, restoring the cord to its original state (right). Slide the key to the right, through the loop, and the job is done.

One reader, Allan Kiron, pointed out that if the cord is long enough, and one is allowed to remove the door from its hinges, the puzzle can be solved by passing a loop of cord over the entire door as if the door were a ring.

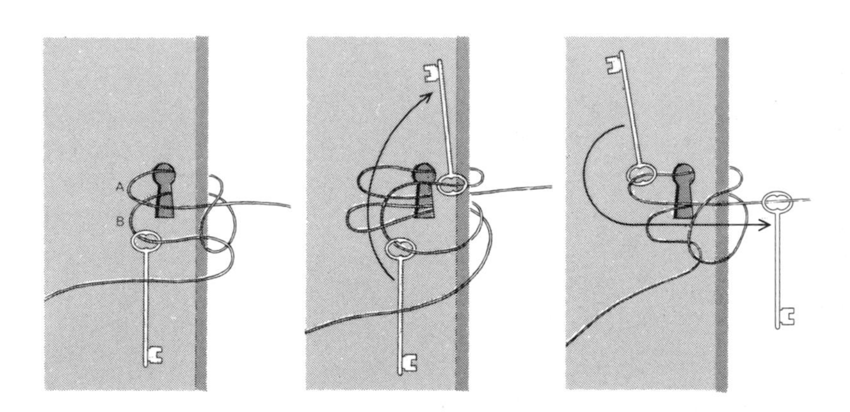

FIGURE 60
Solution to key-and-cord puzzle

4. Dmitri Borgmann, author of *Language on Vacation,* is my authority for the following answers to the questions about the anagram dictionary.

1. Among common words, SU, *us*, is probably the last entry. But this is followed by TTU, *tut,* TTUU,*tutu* (a short ballet skirt), TUX, *tux* (short for tuxedo) and ZZZZ, *zzzz* (to snore), which Borgmann says is in the second edition of *The American Thesaurus of Slang*, by Lester V. Berrey and Melvin Van Den Bark.

2. The first and second entries are A, *a*, and AA, *aa* (a kind of lava). Is the third entry AAAAABBCDRR, *abracadabra?*

3. The last entry is AY, *ay*, unless we accept AYY, *yay* (an obsolete variant of "they").

4. Among common words, the first entry beginning with *B* is probably BBBCDEEOW, *cobwebbed.* It is preceded by the less common BBBBBEEHLLUU, *hubble-bubble* (a bubbling sound, also a hookah).

5. ABCDEFOL, *boldface.*

6. The longest entry that is itself a common word (that is, a word with letters in alphabetical order) is *billowy*. In *Language on Vacation* Borgmann supplies a longer word, *aegilops* (a genus of grasses).

7. The longest common English word that does not repeat a letter is *uncopyrightables.* But Borgmann supplies some longer coined words, such as *vodkathumbscrewingly*, with 20 letters, which means to apply thumbscrews while under the influence of vodka. The longest such word ever created, he says, is the 23-letter monster *pubvexingfjordschmaltzy*, which means "as if in the manner of the extreme sentimentalism generated in some individuals by the sight of a majestic fjord, which sentimentalism is annoying to the clientele of an English inn."

In 1964, shortly after Temperley proposed the compiling of an anagram dictionary, Follett Publishing Company published the *Follett Vest Pocket Anagram Dictionary*, compiled by Charles A. Haertzen. It contains 20,000 words of seven or fewer letters, with an informative introduction and useful bibliography. *Unscrambler*, an anagram dictionary of 13,867 words through seven letters, was published in 1973 by The Computer Puzzle Library, Fort Worth, Tex. An anagram dictionary of more than 3,200 Old Testament names was privately published in 1955 by Lucy H. Love, Darien, Conn. It is titled *Bible Names "De-Koder."* Readers interested in anagrams will enjoy Howard W. Bergerson's monograph, *Palindromes and Anagrams*, a 1973 Dover paperback.

5. It is easy to show that for any finite set of points on the plane there must be an infinity of straight lines that divide the set exactly in half. The following proof for the six points in Figure 61 applies to any finite number of points.

Consider every line determined by every pair of points. Pick a new point, *A*, that lies outside a closed curve surrounding all the other points and that does *not* lie on any of the lines. Draw a line through point *A*. As this line is rotated about point *A*, in the direction shown, it must pass over one point at a time. (It cannot pass two points simultaneously; this would mean that point *A* lay on a line determined by those two points.) After it has passed half of the points inside the curve it will divide the set of points in half. Since *A* can be given an infinity of positions, there is an infinity of such lines.

This problem is based on a quickie problem contributed by Herbert Wills to *The Mathematical Gazette*, May, 1964.

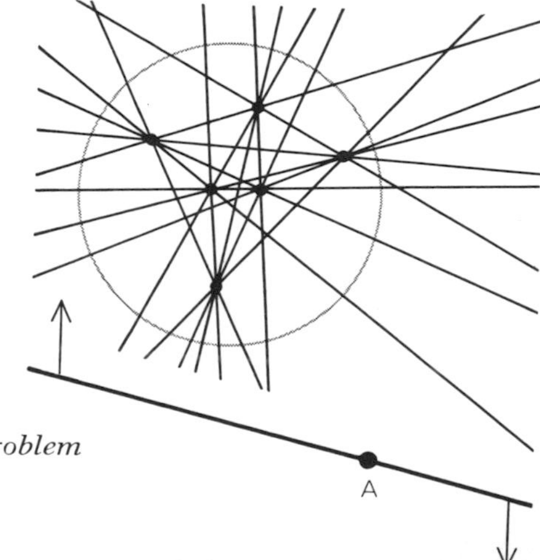

FIGURE 61
Proof for million-point problem

6. If the girl's objective is to escape by reaching the shore as quickly as possible, her best strategy is as follows. First she rows so that the lake's center, marked by the raft, is always between her and the man on shore, the three points maintaining a straight line. At the same time, she moves shoreward. Assuming that the man follows his optimum strategy of always running in the same direction around the lake, with a speed four times as fast as the girl can row, the girl's optimum path is a semicircle with a radius of $r/8$, where r is the lake's radius. At the end of this semicircle, she will have reached a distance of $r/4$ from the lake's center. That is the point at which the angular velocity she must maintain to keep the man opposite her just equals his angular velocity, leaving her no reserve energy for moving outward. (If during this period the man should change direction, she can do as well or better by mirror reflecting her path.)

As soon as the girl reaches the end of the semicircle, she heads straight for the nearest spot on the shore. She has a distance of $3r/4$ to go. He has to travel a distance of pi times r to catch her when she lands. She escapes, because when she reaches the shore he has gone a distance of only $3r$.

Suppose, however, the girl prefers to reach the shore not as *soon* as possible but at a spot as *far away* as possible from the man. In this case her best strategy, after she reaches a point $r/4$ from the lake's center, is to row in a straight line that is tangent to the circle of radius $r/4$, moving in a direction opposite to the way the man is running. This was first explained by Richard K. Guy in his article, "The Jewel Thief," in *NABLA*, Vol. 8, September, 1961, pages 149–150. (This periodical, a bulletin of the Malayan Mathematical Society, had published the minimum-time solution of the problem in its July, 1961, issue, page 112.)

Using elementary calculus, Guy shows that the girl can always escape even when the man runs 4.6+ as fast as the girl rows. The same results are given by Thomas H. O'Beirne in *The New Scientist*, No. 266, December 21, 1961, page 753, and by W. Schurman and J. Lodder in "The Beauty, the Beast, and the Pond," *Mathematics Magazine*, Vol. 47, March, 1974, pages 93–95.

7. The smallest number of unit line segments that can be removed from a four-by-four checkerboard to render it "square-free" is nine. One way to do this is shown on the interior four-by-four square at the top of Figure 62.

To prove this minimum, note that the eight shaded cells have no side in common; to break the perimeters of all eight, at least eight unit lines must be removed. The same argument applies to the eight white cells. We can, however, "kill" all 16 cells with the same eight lines if we pick lines shared by adjacent cells so that each erased line kills a white and a shaded cell simultaneously. But if we do this, not one of the removed lines will be on the board's outside border, which forms the largest square. Therefore at least nine cells must be removed to kill the 16 small cells plus the outer border. As the solution shows, the same nine will eliminate all 30 squares on the board.

The same argument proves that every even-order square must have a solution at least equal to $\frac{1}{2}n^2 + 1$, n being the square's order. Can this be achieved on all even-order squares? A proof by induction is implicit in the procedure shown in the illustration. We merely plug a domino in the open cell on the border of the four-by-four, then run a chain of dominoes around the border as shown. This provides a minimum solution of 19 for the order-six board. The same procedure is applied once more to give the minimum solution of 33 for the eight-by-eight board. It is obvious that this procedure can be repeated endlessly, with each new border of dominoes raising the open cell one step, as shown by the arrow.

On the order-five board the situation is complicated by the fact that there is one more shaded cell than there are white cells. At least 12 lines would have to be removed to kill 12 shaded and 12 white cells simultaneously. This would, of course, form 12 dominoes. If the remaining shaded cell were on the outside border, both this cell and the border could be killed by taking one more line, which suggests that odd-order squares might have a minimum solution of $\frac{1}{2}(n^2 + 1)$. To achieve this, however, the dominoes would have to be arranged so as not to form an un-

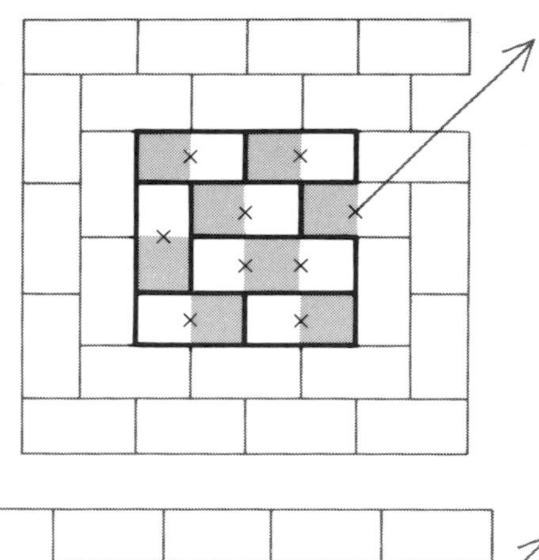

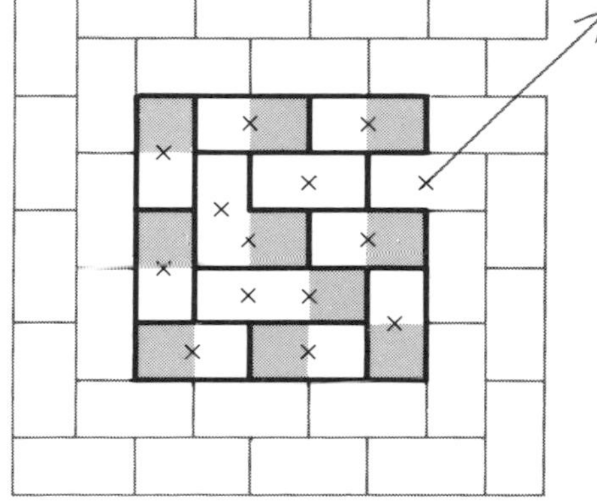

FIGURE 62
Solutions for toothpick problems

broken square higher than order-one. It can be shown that this is never possible, so that the minimum is raised to $\frac{1}{2}(n^2 + 1) + 1$. The lower drawing in Figure 62 shows a procedure that achieves this minimum for all odd-order squares.

D. J. Allen, George Brewster, John Dickson, John W. Harris, and Andrew Ungar were the first readers to draw a single diagram that could be extended to display all solutions rather than separate diagrams as I had found, for the odd and even cases.

David Bienenfeld, John W. Harris, Matthew Hodgart, and William Knowlton, attacking the companion problem of creating rectangle-free patterns, discovered that the L-tromino plays in this problem the same role the domino plays in the square-free problem. For squares of orders 2 through 12, the minimum number of lines that must be removed to kill all rectangles are, respectively: 3, 7, 11, 18, 25, 34, 43, 55, 67, 82, 97. Perhaps at some future time I can comment on the formulas and algorithms that were developed. Figure 63 shows a pattern for order-eight.

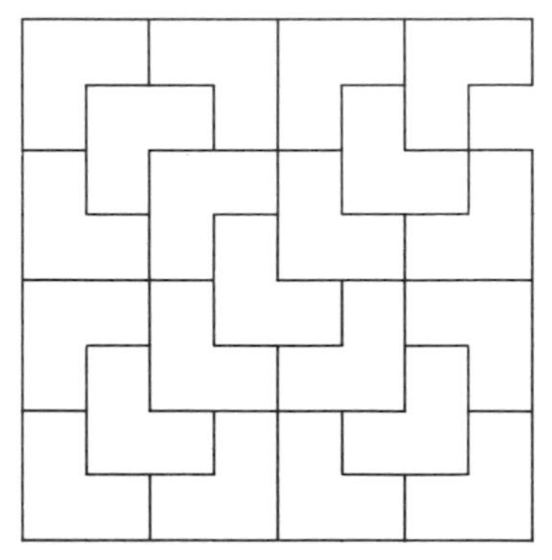

FIGURE 63
Rectangle killing on the order-eight lattice

8. Three sets of four cocircular points on the picture of randomly dropped rectangles and disks are shown as black spots in Figure 64. The four corners of the rectangle were mentioned in the statement of the problem. The four points on the small circle are obviously cocircular. The third set consists of points A, B, C, D. To see this, draw dotted line BD and think of it as the diameter of a circle. Since the angles at A and C are right angles, we know (from a familiar theorem of plane geometry) that A and C must lie on the circle of which BD is the diameter.

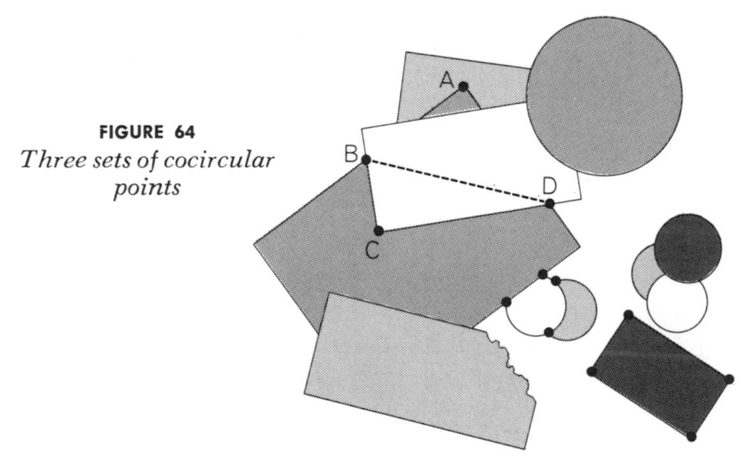

FIGURE 64
Three sets of cocircular points

When this problem was first published, I asked only for three sets of four cocircular points. The problem proved to be better than either its originator or I realized. Many readers were quick to call attention to the fourth set. Its four points are: *A*, the unmarked intersection immediately below *A* on the right, *B*, and the unmarked corner just above *B*. The line segment joining *B* to the point below *A* is the diameter of a circle on which the other two points lie, since each is the vertex of a right angle subtending the diameter.

The problem, modified to exclude this fourth set of points, appears in Barr's *Second Miscellany of Puzzles* (Macmillan, 1969).

9. This is how I originally answered the problem:

The most efficient procedure for testing any number of glasses of liquid in order to identify a single glass containing poison is a binary procedure. The glasses are divided as nearly in half as possible. One set is tested (by mixing samples from all the glasses and testing the mixture). The set known to include the poison glass is then again divided as nearly in half as possible, and the procedure is repeated until the poison glass is identified. If the number of glasses is from 100 to 128 inclusive, as many as seven tests might be required. From 129 to 200 glasses might take eight tests. The number 128 is the turning point, because it is the only number between 100 and 200 that is in the doubling series: 1, 2, 4, 8, 16, 32, 64, 128, 256. . . . There must have been 129 glasses in the hotel kitchen, because only in that case (we were told that the number was between 100 and 200) will the initial testing of one glass make no difference in applying the most efficient testing procedure. Testing 129 glasses, by halving, could demand eight tests. But if a single glass is tested first, the remaining 128 glasses require no more than seven tests, so that the total number of tests remains the same.

When the above answer appeared, many readers pointed out that the police commissioner was right, and the mathematician wrong. Regardless of the number of glasses, the most efficient testing procedure is to divide them as nearly in half as possible at each step and test the glasses in either set. When the probabilities are worked out, the expected number of tests of 129 glasses, if the halving procedure is followed, is 7.0155+. But if a single glass is tested first, the expected number is 7.9457+. This is a rise of .930+ test, so the commissioner was almost right in considering the mathematician's procedure a waste of one test. Only if there had been 129 glasses, however, do we have a plausible excuse for the error, so, in a way, the problem was correctly answered even by those readers who proved that the mathematician's test procedure was inefficient.

The problem appears in Barr's *Second Miscellany of Puzzles*.

CHAPTER 10

Card Shuffles

Shuffling is the only thing which Nature cannot undo.
—Sir Arthur Eddington,
The Nature of the Physical World

Eddington was probably wrong. In 1964 physicists discovered that certain events, involving the weak interaction of fundamental particles appear not to be time-reversible. It seems likely that nature does these things in only one time direction, unless there are galaxies or regions of the cosmos where matter is not only reflected and charge-reversed (that is, where it is antimatter) but also moving in a time direction opposite to our own. No one knows yet what connection all this has, if any, with the macroworld in which shuffling processes provide the only physical basis for what Eddington called the "arrow of time."

Apart from the newly discovered anomalies, all fundamental laws of physics, including the laws of quantum physics, are time-reversible. You can change the sign in front of t from plus to minus, and the formula describes something nature can do. But when a large number of objects, from molecules to stars, are moving randomly, the statistical laws of probability introduce the time arrow. If gas A and gas B are in the same container but separated by a partition, and the partition is removed, the molecules of the two gases shuffle together until the mixture is homogeneous. It never unshuffles. As far as individual molecules are concerned, there is no reason why each could not be given a direction and velocity that would "undo" the mixture. It doesn't happen because the probability of such a sorting is virtually zero. Here, Eddington argued (and most physicists agreed), is the only reason why a dropped egg never puts itself together again and hops back up to the edge of the wall. Probability laws decree that the billions of molecules that scatter randomly during such an event will move so as to increase the entropy (the measure of a certain kind of disorder) of the total system. The universe, prodded by probability, shuffles along the time axis in one direction only.

The shuffling of a deck of cards is, as Eddington pointed out, a splendid paradigm of Nature's one-way shuffle habits. Arrange a deck so that the top 26 cards are red, the bottom 26 black. The situation is analogous to the container of two gases. Shuffle the deck 10 times and the red-black order is obliterated. Why is it that continued shuffling does not sort the deck back into red-black halves? Because there are 52! different ways a deck can be arranged. (The exclamation mark is the factorial sign indicating the product of $1 \times 2 \times 3 \times 4 \times \ldots$ and so on up to 52. It is a number of 68 digits beginning with 8.) Of these 52! permutations, the number that exhibit a complete red-black separation, although very large, still constitute such a small fraction of 52! that one could shuffle for thousands of years without expecting to hit a single one of them.

The curious thing about shuffling cards is that a shuffle's efficiency—its power to introduce randomness into an ordered deck—actually depends on the clumsiness of one's fingers. Unless the cards drop in a disorderly way the shuffle doesn't really shuffle. Consider, for example, the "overhand" shuffle. The deck is held by its ends, in the right hand, and the left thumb "milks" the cards off the top in small packets of random size. A perfect overhand shuffle, in which the thumb takes one card at a time, does not destroy the order of the deck at all. It just reverses it. A second perfect overhand shuffle restores the original order.

The more familiar "riffle" shuffle, performed on a table, also fails to do its job if done perfectly. The perfect riffle shuffle, known to American magicians as the "faro shuffle" and to English magicians as the "weave shuffle," is one in which the cards drop one at a time, and alternately, from the two thumbs. The deck must, if it contains an even number of cards, be divided exactly in half before the shuffle begins, and as nearly in half as possible if it contains an odd number of cards. With odd decks the smaller half (one card fewer) shuffles into the larger one so that the top and bottom cards of the larger half become the top and bottom cards of the deck after the shuffle is completed. With even decks you have a choice of dropping first the bottom card of either half. If the first card to fall is from the half that was formerly the bottom of the deck, the cards previously at the top and bottom will remain at the top and bottom. Magicians call this the "out-shuffle" because the top and bottom cards remain on the outside. If the first card to fall is from what was formerly the top half of the deck, the former top and bottom cards go *into* the deck to positions second from top and bottom. Magicians call this the "in-shuffle."

For odd decks, a faro is an out-shuffle if it is cut below the center card. This places the top card on the larger half, with the result that it remains the deck's top card after the shuffle. The faro is an in-shuffle if it is cut above the center card. This places the top card on the smaller half, with the result that it becomes the second card after the shuffle. Both in- and out-shuffles of odd decks are called straddle shuffles (the larger half "straddles" the smaller half), a term coined by Ed Marlo, a Chicago card expert who has written several books on the faro and invented many elegant card tricks which depend on the faro.

A deck of n cards, given a repeated series of faro shuffles of the same type, will return to its original order after a finite number of shuffles. If n is odd, the deck returns to its initial state after x shuffles, where x is the exponent of 2 in the formula $2^x = 1$ (modulo n). "1 (modulo n)" means that the number has a remainder of 1 when it is divided by n. For example, if a joker is added to a full deck, making 53 cards, the formula becomes $2^x = 1$ (modulo 53). We must find an integral value of x such

Number of Cards in Deck	Number of Faro Shuffles Required to Restore Order	
	Out-Shuffles	In-Shuffles
2	1	2
3	2	2
4	2	4
5	4	4
6	4	3
7	3	3
8	3	6
9	6	6
10	6	10
11	10	10
12	10	12
13	12	12
14	12	4
15	4	4
16	4	8
17	8	8
18	8	18
19	18	18
20	18	6
21	6	6
22	6	11
23	11	11
24	11	20
25	20	20
26	20	18
27	18	18

Number of Cards in Deck	Number of Faro Shuffles Required to Restore Order	
	Out-Shuffles	In-Shuffles
28	18	28
29	28	28
30	28	5
31	5	5
32	5	10
33	10	10
34	10	12
35	12	12
36	12	36
37	36	36
38	36	12
39	12	12
40	12	20
41	20	20
42	20	14
43	14	14
44	14	12
45	12	12
46	12	23
47	23	23
48	23	21
49	21	21
50	21	8
51	8	8
52	8	52

FIGURE 65

The number of shuffles required to restore order to a deck of from two to 52 cards

that 2^x has a remainder of 1 when divided by 53. If we go up the ladder of the powers of 2 (2, 4, 8, 16, 32 . . .), we do not reach a number that is 1 (modulo 53) until we come to 2^{52}. This tells us that 52 in-shuffles (or 52 out-shuffles) are required to restore the order of a 53-card deck.

If the deck is even, the situation is a bit more complicated. The number of out-shuffles needed to restore the original order is $2^x = 1$ [modulo $(n - 1)$]. The number of in-shuffles that does the trick is $2^x = 1$ [modulo $(n + 1)$]. This sometimes makes a big difference. For the normal pack of 52 cards, 52 in-shuffles restore order. But $2^8 = 1$ (modulo 51), so that only eight out-shuffles are needed!

Figure 65 gives the number of faro shuffles of both types required to restore the order of a deck of any size from two to 52 cards. Note that for an odd deck the number is always the same for either type of shuffle, and equal to the number of out-shuffles required for a deck of one more card. For an even deck the number of out-shuffles is the same as the number of in-shuffles for a deck of two cards fewer. This reflects the fact that the top and bottom cards are never disturbed during out-shuffles and so you are in effect merely in-shuffling the rest of the deck.

Since it is difficult to perform perfect riffle shuffles even clumsily (only a skilled card expert can simulate a genuine shuffle), we can best test the accuracy of this chart by reversing time and doing faro shuffles backward. (*Perfect* shuffles are easy to undo!) Card magicians call this maneuver a "reverse faro." Simply fan through a deck as shown in Figure 66, jogging alternate cards up out of the deck (dotted lines in top drawing). With practice this can be done rapidly. After all the cards have been jogged apart, strip the half-decks apart and put one on the other. If you replace the cards so that the top card remains on top, you have performed an "out-sort." If the top card goes into the deck, you have done an "in-sort." Each operation is obviously the inverse of the corresponding faro. It is now a simple matter to test any part of the chart, for if n faros of a certain type restore an order, then n reverse faros clearly will do the same thing.

FIGURE 66
Technique of the "reverse faro" shuffle

It is best to experiment with ordered sets of cards that are held face up so that you can see how the pattern shifts with each sort. Observe, for example, that in certain cases where an even number of sorts restore the order, the cards become arranged in reverse order after half of the sorts are completed. Try 10 cards with values from ace to 10, and in serial order. Ten in-sorts restore the original order, but after five in-sorts the deck is in reverse serial order. A deck of 52 cards similarly reverses its order after 26 in-sorts. Note also the curious fact that each time an in-sort is made with the 52-card deck the 18th and 35th cards trade places.

Alex Elmsley, a British computer programmer who is also a skilled card magician, was one of the first to explore the intricate mathematics of the faro from a conjurer's point of view. Writing in 1957 in *Ibidem,* a Canadian magic journal, he told how he had hit on a remarkable formula. He had earlier coined the terms "in-shuffle" and "out-shuffle," and in his notes had been abbreviating them with I (for "in") and O (for "out"). One of his first problems was to determine what sequence of in- and out-shuffles would be the most efficient in causing the top card of a deck to go to any desired position from the top. For example, suppose a magician, using a full deck, wishes to faro-shuffle the top card to the 15th position. Elmsley found experimentally that this could be done by the following sequence of faros: IIIO. This, he recognized at once, is also the way to write 14 in binary notation, and 14 is the number of cards above the desired position!

It was no coincidence. Regardless of the size of a deck, or whether it is odd or even, the following procedure always works. Subtract 1 from the position to which you want to bring the top card. Express the result as a binary number and you have the proper sequence of in- and out-shuffles to put the card there in the shortest possible time.

If the deck is even, there is an unexpected bonus. Whenever a faro is made, the downward shifts of cards in the top half are exactly mirrored by upward shifts of the bottom cards. While the card that is *n*th from the top goes to position *p* from the top, the card that is *n*th from the bottom goes to position *p* from the bottom. The same shuffle that puts the top card 15 down will simultaneously bring the bottom card 15 up. If a deck of 52 cards is arranged so that each card in the top half matches in value and color the card in the same position from the bottom, this matching is never destroyed by any number of faros, of either type. Magicians know it as the "stay-stak" principle, a term coined by its discoverer, a card magician who wrote under the name of Rusduck. Many brilliant card effects are based on this mirroring principle.

The reader may enjoy testing Elmsley's formula by performing another time reversal, using in- and out-sorts to bring a card from any position, in any size of deck, to the top. Subtract 1 from the position number, write the result in binary form, then follow the sequence of binary digits *backward*, doing in- or out-sorts as indicated. In the previous example, to bring the 15th card to the top, write 14 as 1110. A sequence of one out-sort followed by three in-sorts puts the card on top. If the deck is even, the 15th card from the bottom simultaneously goes to the bottom of the deck.

The mirror principle does not apply to odd decks, but something even more astonishing obtains. It can best be understood by experimenting with a nine-card packet containing the ace, 2, 3, 4, 5, 6, 7, 8, and 9 of one suit. Arrange these cards in serial order, all face up, ace on top. Think of this as a cyclic order, the top and bottom cards joining like a closed chain. If you cut the packet to produce, say, the order 6, 7, 8, 9, 1, 2, 3, 4, 5, we shall call this the *same* cyclic order. The chart shows that a packet of nine cards returns to its original order after six sorts of either type. Now, however, you intersperse the sorts with as many cuts of the packet as you please. Cut one or more times, do a sort, cut some more times, do another sort, and so on until you complete the six sorts. Moreover, you can mix in- and out-sorts as you please. After the sixth sort examine the cards. They will be in the same cyclic order! This applies to any odd deck. Sorts of either type can be mixed with any number of cuts, and after the required number of sorts the original cyclic order is restored.

While the nine-card deck is being restored to its original state, it goes through five other states, each with its own cyclic order. The cyclic orders of these other states are also undisturbed by the cutting; they simply show up in different cyclic permutations, depending on which card is on top. Since each of the deck's six states has only nine different cyclic permutations, it follows that the total number of permutations of the nine cards that can be obtained by mixing cuts with sorts is no more than $6 \times 9 = 54$. This is only a small fraction of the $9! = 362{,}880$ possible permutations of nine cards.

Because sorts are time reversals of faros, all of this applies equally to the faro shuffling of odd decks. Solomon W. Golomb, in his paper on "Permutations by Cutting and Shuffling," proved that by mixing cuts with in-shuffles or with out-shuffles an *even* deck could reach any one of its possible permutations. But for all *odd* decks of more than three cards only a small portion of the possible permutations are obtainable. Random cutting and faro shuffling can induce complete randomness in a deck of 52 cards because every one of the 52! possible permutations can be reached. But remove one card from the deck, leaving 51 cards, and no amount or mixture of faro shuffles and cuts will ever yield more than $8 \times 51 = 408$ permutations, out of the total of 51! possible permutations—a number of 67 digits.

Dai Vernon, one of the nation's top card magicians, has based an easy-to-do trick on this cyclic character of odd decks. Hand 20 cards to someone, and a joker. Ask him to shuffle the cards while your back is turned, insert the joker into the packet and remember the two cards the joker goes between. Turn around and take from him the packet of 21 cards, all face down. Do a reverse faro (either an in- or an out-sort, it makes no difference), then ask him to cut the packet. Repeat with another reverse faro, of either type, and let him cut again. Now spread the cards into a fan and hold the fan up so that the spectator can see the faces but you cannot. Ask him to remove the joker.

Break the cards into two groups at the spot occupied by the joker, then put them together again the other way. In other words, you cut the packet at the point that was occupied by the joker. Do not call attention to this, however. To the audience it should look as if you merely put the cards together again as they were before the joker was removed.

You now hold 20 cards, an even number. Do two out-sorts and one in-sort. Put the packet on the table. Ask the spectator to name the two selected cards. Show the bottom card of the packet. It will be one of the cards. Turn over the top card of the packet. It will be the other card.

The faro is only one of many types of simple, strongly patterned shuffles that can be applied repeatedly to a deck with unusual results. Let us now generalize and define a "shuffle" as any transformation whatever, patterned or nonpatterned. We can specify the structure of the shuffle by writing a table such as the following one for a five-card shuffle:

1–3
2–5
3–1
4–2
5–4

The table shows that the first card goes to position 3, the second card to position 5, and so on. The same shuffle is diagrammed with arrows in Figure 67. There need be no pattern of any kind in the placement of these arrows. The pattern can be completely random, as if the cards were shaken in a barrel, then taken out one at a time to form a new deck.

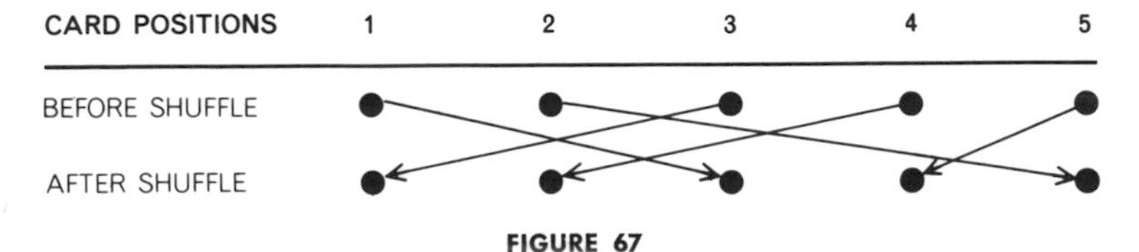

FIGURE 67

Diagram for a randomly patterned shuffle of five cards. It has a cycle of six shuffles.

Assume that exactly the same shuffle, as specified by a table or diagram, is repeatedly applied to a deck of *n* cards. Will this eventually randomize the cards? No, it will not. Regardless of the shuffle pattern, the cards simply progress through a series of states, no two alike, until they return to their original order, then the cycle repeats. If the deck contains more than two cards, there is no shuffle that repeated will run through all possible permutations. Three cards, for instance, have 3!, or $1 \times 2 \times 3 = 6$, possible orderings. It is impossible to devise a shuffle that, if it is repeated, will require six steps to complete its cycle. The longest cycle possible is three steps.

This suggests a difficult but fascinating question. Imagine that a deck of 52 cards has been put inside a shuffling machine that keeps repeating exactly the same shuffle. You cannot see into the machine and so you have no notion of the shuffle pattern. Each time a shuffle is completed a bell rings. What is the smallest number of rings after which you can say with absolute certainty that the original order has returned at least once? Put another way: What is the longest cycle a repeated shuffle of 52 cards can have?

ADDENDUM

IN DISCUSSING the repeated pattern shuffle, I stated without proof that such a shuffle was sure to return the deck to its original order after a finite number of shuffles. Several readers wondered why it was not possible for such a shuffle to enter a "loop" that would never return to the original order.

Here is why this cannot happen. When a deck is repeatedly given the same shuffle, it passes through a series of states: *a,b, c,d,e,* . . . . When the shuffle is applied, say, to *b*, it must produce *c*. Conversely, *c* can be produced only by applying the shuffle to *b*. Since the deck has a finite number of states, it must return to its original state unless somewhere along the line it returns to a state other than *a*. Clearly it cannot do this. It cannot, for instance, return to *d* without returning first to *c* (because *d* is produced only from c), and it cannot return to c without returning first to b, and it cannot return to b without returning to a. The chain loops only by returning to a.

The same reasoning applies to any packet of cards when given a series of n shuffles, each a different pattern, provided the series is exactly repeated. Suppose a deck is given three different shuffles, a,b,c, and this is repeated: abc, abc, abc, ... Each triplet changes the deck the same way from one state to another, so the triplet is equivalent to a single shuffle.

It is also easy to show that for every x, where x is the smallest number of repeated faros of the same type required to restore the original order, only a finite number of decks are restored by x shuffles. For example: only decks of 5, 6, 14, 15, and 16 cards require a minimum of four shuffles to restore them (four out-shuffles for 5, 6, 15, 16, and four in-shuffles for 14). Readers may enjoy working out a procedure for determining all deck sizes that can be restored with a minimum of x faros of the same type.

One of the earliest mentions of the faro shuffle is in John Nevil Maskelyne's book on card cheating, *Sharps and Flats* (1894), where it is called the "faro dealer's shuffle." Charles T. Jordan, in *Thirty Card Mysteries* (1919), was the first magician to give serious thought to how the shuffle could be applied to card tricks. It was not until the late fifties, however, that card magicians began to master the shuffle in earnest and explore its possibilities in depth. "The soft whir of perfectly faro'ed cards interweaving alternately is to be heard throughout magicdom," wrote John Braun in his introduction to Paul Swinford's *Faro Fantasy* (1968).

At every gathering of card magicians you will find the faro enthusiasts anxious to show their latest creations. And you will also find card men who, although they may be able to do excellent faros, avoid all faro tricks whenever they are entertaining laymen. "A friend of mine picked up a deck of cards and said he was going to show me a faro trick," wrote top card magician Charlie Miller. "I took out a gun and shot him."

Many unexpected aspects of the faro have been discovered by magicians. Let me cite only one instance. Take a packet of 32 cards (any power of 2 will work), with the four aces included. Put one black ace on top, the other on bottom. Put one red ace at position n from top, the other red ace at position n from the bottom. Let's say $n = 7$. Turn face up the ace that is seventh from top. Now do five reverse faros, each time stripping the half with the reversed ace to the top. Be sure to complete all five, even if the reversed ace comes to the top in fewer faros. At the finish you will have the red aces on top and bottom, and the black aces will have moved to the seventh positions from top and bottom!

Dozens of excellent card tricks exploit this principle. For example, ask someone to shuffle a packet of 16 cards and hand it to you. Secretly glimpse the top card and remember it. Fan the packet so the spectator can see the faces. Ask him to think of any card, remember it, and also remember its position from the top. Do four reverse faros, holding the cards so he can see the faces after you jog them. Before each stripping, ask him to tell you which half his card is in. Strip that half to the top. At the finish, ask for the name of his card, and turn over the top card to show that the sorting procedure has brought his card to the top.

Offer to repeat the trick, but this time you will ask him for *no information whatever*. Turn your back while he looks at the card now in the position formerly occupied by his previously selected card. Of course you already know the name of this card, which you had secretly glimpsed when it was on top, but he has no reason to suspect that you know it. Take back the packet and do four reverse faros, this time with the faces toward you and without asking any questions. Each time, strip the packet with his card to the top. Square the cards, ask for the name of his second chosen card, turn over the top card. He told you nothing, yet the trick worked as well as before.

My earlier remark that the mirror (or stay-stak) principle does not apply to odd decks is not strictly true. Edward Marlo wrote to point out that a 53-card deck, with the joker on top or bottom, can be straddle-faroed as often as you please (using either type of straddle faro), and cut as often you like between shuffles. If the joker is now cut back to top or bottom, you will find that the deck, ignoring the joker, has retained the mirroring. Marlo, Charles Hudson, and other card men have designed many unusual card tricks exploiting the mirror principle with a 53-card deck.

I explained how the faro could be used to bring the top card to any desired position, and how the reverse faro could be used to bring a card from any position to the top. The reverse problem of bringing a card from any position to the top with standard faros (or the corresponding task of using reverse faros to put the top card in a desired position) is much harder to analyze. So far as I know, no simple formulas or procedures have been found for doing this with a minimum number of shuffles. Some attempts at efficient algorithms, combining shuffles of different types, have been proposed, but the problem is far from satisfactorily disposed of.

Readers who care to venture into the jungle of faro card magic will find the major references listed in the bibliography.

ANSWERS

ASSUMING THAT a shuffle pattern is repeated exactly each time, what is the maximum number of iterations that can be made before a 52-card deck returns to its original order?

This is best answered by first considering the random shuffle of six cards diagrammed in Figure 68. Examine it closely and you will see that it can be broken into subsets, each with its individual cycle. The card in position 3 goes to position 3, and so it forms a subset of one card with a cycle of 1. Cards 1 and 5 interchange, forming a subset of two cards that return to their original positions after a two-shuffle cycle. Cards 2, 4, and 6 are in a subset that returns to the initial order after three shuffles. We thus have three cycles of lengths 1, 2, and 3. It is obvious

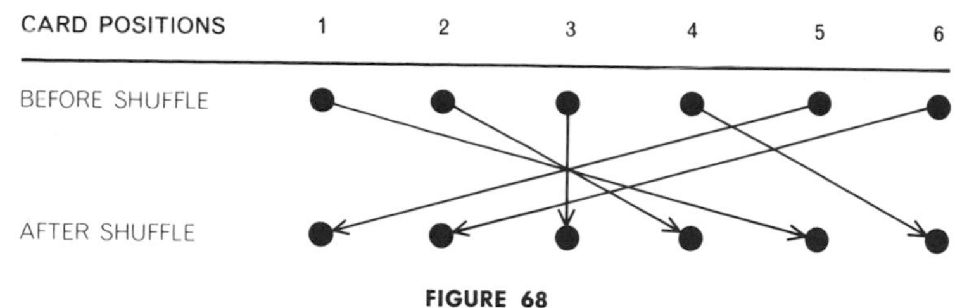

FIGURE 68

Diagram of a random shuffle of six cards

that the entire deck of six cards will return to its initial order after a number of shuffles equal to the LCM (lowest common multiple) of 1, 2, and 3, which is 6. All decks, when a shuffle is iterated, subdivide into such subsets, each of which has a cycle equal to its number of cards. To find the longest cycle for an entire deck of n cards, we test every possible partition of n into subsets to see which partition gives sets with the *highest* LCM. In the case of six cards there are 11 different partitions:

1 1 1 1 1 1
1 1 1 1 2
1 1 2 2
2 2 2
1 1 1 3
1 2 3
3 3
1 1 4
2 4
1 5
6

The subsets with the highest LCM's are the 1, 2, 3 set and the 6 set, both of which have 6 as the LCM. We conclude that no shuffle of six cards, exactly repeated, can have a cycle longer than 6 before the original order is restored.

A deck of 52 cards has so many different partitions that one must use shortcuts to find those partitions whose sets have the highest LCM. There is not space to go into this here; I can only refer the reader to W. H. H. Hudson's article in the *Educational Times Reprints: Volume II* (1865), page 105, where the problem seems first to have been solved. No partition of 52 has an LCM higher than 180,180; therefore no shuffle for 52 cards can have a cycle longer than 180,180. An example of such a partition is 1, 1, 1, 4, 5, 7, 9, 11, 13. The reader should have no difficulty diagramming a 52-card shuffle, with subsets corresponding to the numbers in the partition, that will not return a deck to its original order until the shuffle has been repeated 180,180 times.

The curious repetition of 180 is explained by the fact that partitions 7, 11, and 13 have a product of 1,001, and the remaining partitions have a product of 180. Any three-digit number *abc* multiplied by 1,001 has a product of *abcabc*. The single-card subsets obviously play no role in the shuffle. We conclude, therefore, that decks of 49, 50, or 51 cards also have maximum shuffle cycles of 180,180 repetitions. Adding a joker to the deck raises the maximum cycle to 360,360. For a recent discussion of the problem, which cites earlier references, see "Lost in the Shuffle," solution to Problem E2318, *American Mathematical Monthly*, Vol. 79, October, 1972, page 912.

Instead of asking for the maximum length of a cycle for 52 cards, suppose we ask a different question. Assume that the machine cannot be put into reverse to "undo" a shuffle. We have given it a deck of 52 cards, in an unknown order, and the machine has given it one shuffle, the nature of which is also unknown. How many additional shuffles of the same pattern must we have the machine make in order to be certain we have restored the deck's original order?

Two readers, Edwin M. McMillan and Daniel Van Arsdale, independently asked and answered this question. The smallest cycle guaranteed to restore the original order is the least common multiple of all numbers from 1 through 52. This number is $11 \times 13 \times 17 \times 19 \times 23 \times 25 \times 27 \times 29 \times 31 \times 32 \times 37 \times 41 \times 43 \times 47 \times 49$. Call this very large number N. If the machine now repeats its shuffle $N - 1$ times, we can be certain that the deck has returned to its original state.

CHAPTER 11

Mrs. Perkins' Quilt and Other Square-Packing Problems

THE MATHEMATICAL and physical sciences edition of *The Proceedings of the Cambridge Philosophical Society*, one of the least frivolous of British journals, startled its readers in July, 1964, by publishing a lead article that bore the title "Mrs. Perkins' Quilt." It was a technical discussion by the Cambridge mathematician J. H. Conway (see Chapter 1) of one of the most useless but intriguing unsolved problems in recreational geometry.

The problem belongs to a large family of combinatorial questions that involve the packing of squares into larger squares. The best-known problem of this type is that of fitting a set of squares no two of which are alike into a larger square without any overlap or leftover space. If we think of the larger square as a lattice of unit squares to be divided along lattice lines into unequal squares, the smallest-known square that can be so divided has a side of 175 units. It can be cut into 24 unequal squares. The reader will find a picture of it on page 206 of *The 2nd Scientific American Book of Mathematical Puzzles & Diversions*, in a chapter by William T. Tutte explaining how he and his friends used electrical-network theory to find "squared squares" of this type.

The problem of Mrs. Perkins' quilt (it was named by the English puzzlist Henry Ernest Dudeney when he first introduced it) is the same as the problem considered by Tutte except for the elimination of one constraint: the smaller squares *need not be different.* A square lattice of any order n obviously can be divided into n^2 unit squares. The problem, however, is to determine the smallest number of squares into which it can be divided. This seems like a less constrained version of the Tutte problem, but the relaxed conditions do not appear to make the analysis any easier.

Mrs. Perkins' quilt is best approached by starting with the smallest sizes [*see Figure 69*]. Solutions for squares of order 1 and order 2 are trivial. The order-3 square has the unique six-square pattern shown. (Rotations and reflections are not considered different.) Because 4 is a multiple of 2, the order-4 square can be divided, like the order-2, into four equal squares. But since this is merely a blown-up version of the order-2 pattern, we add a new proviso: the smaller squares must not have a common divisor. This leads to the minimum seven-square pattern shown, the minimum pattern that cannot be drawn on a lower-order square. Such a dissection is called a "prime dissection" of the square. Any solution for a square whose side is a prime will be a prime dissection, but for nonprime-order squares we must make sure that the dissection is prime, otherwise the minimum pattern will simply be a trivial repetition of the minimum pattern for the square whose order is the lowest factor of the square's side. Mrs. Perkins' problem can now be stated precisely as that of finding minimum prime dissections for squares of any order. Solutions for the first 12 squares are shown in Figure 69.

When the square's order is in the Fibonacci series 1, 1, 2, 3, 5, 8, 13 . . . (in which each term is the sum of the preceding two terms), a minimal *symmetrical* prime dissection is obtained by dividing it into squares with sides in the Fibonacci series. This produces minimum patterns for orders 1, 2, 3, 5, and 8, as shown, but breaks down for the order-13. Figure 70 shows a symmetrical Fibonacci dissection for order-13. Readers are in-

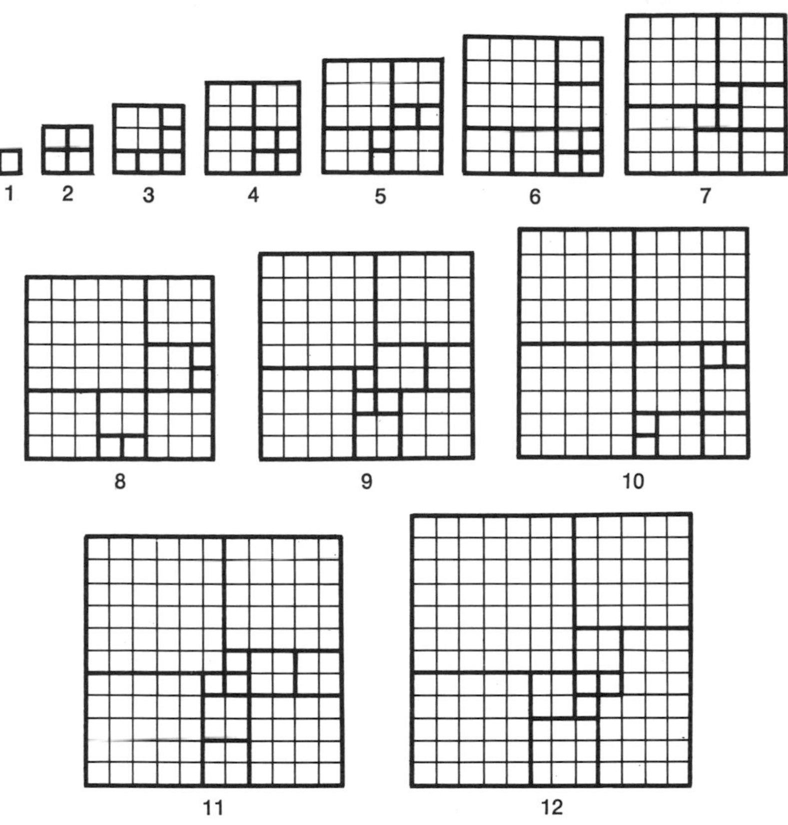

FIGURE 69
Solutions to the quilt problem for the first 12 squares

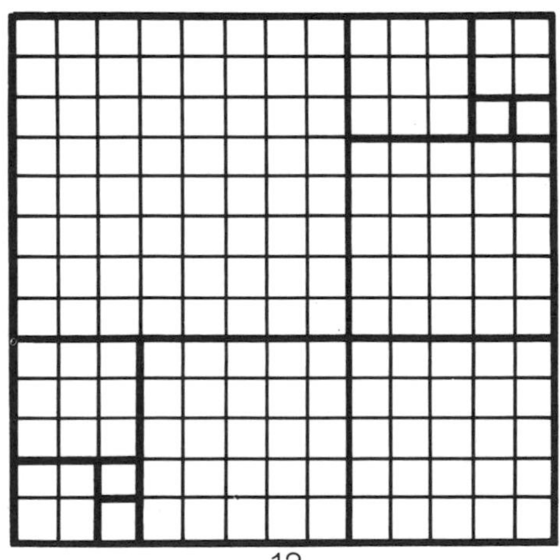

FIGURE 70
Symmetrical pattern for order-13 square

vited to see if they can reduce this from 12 to 11 squares, the minimum, by departing from symmetry; that is, by producing a pattern not superposable on its mirror image.

What is desired, of course, is a general procedure by which minimum prime dissections can be found for squares of any order, and a formula that expresses the minimum number of squares as a function of the order of the larger square. Answers to both questions are nowhere in sight. Conway proved that the minimum prime dissection for a square of order n was equal to or greater than $\log_2 n$, and equal to or less than $6 \sqrt[3]{n} + 1$. In 1965 G. B. Trustrum, of the University of Sussex, published a proof that the least upper bound is $6 \log_2 n$. For higher-order squares this is an improvement over Conway's result, but it is still far from an explicit formula.

Leo Moser, who was head of the mathematics department at the University of Alberta, is cited in Conway's article for his early work on Mrs. Perkins' quilt. In his later years Moser turned to several other square-packing problems. Consider, for example, squares with sides that form the harmonic series $1/2 + 1/3 + 1/4 + 1/5. \ . \ . \ .$ The sum of these sides increases without limit. But the *areas* of these squares form a different series, $1/4 + 1/9 + 1/16 + 1/25 \ . \ . \ .$, that converges, surprisingly on the limit $(\pi^2/6) - 1$ (surprising because of that unexpected appearance of pi). This is a little more than .6. Moser first asked himself: Can this infinite set of squares be fitted, without overlap, inside a unit square? The answer is yes. Figure 71 shows his simple way of doing it. The square is first divided into strips with widths of 1/2, 1/4, 1/8. . . . Because this series has the limit sum of 1, an infinity of such strips can be placed inside the unit square. Within each strip, squares are placed in descending order of size, starting on the left with a square that fills that end of the strip. In this way the infinite set of squares is comfortably accommodated, with a trifle less than .4 of the large square remaining uncovered.

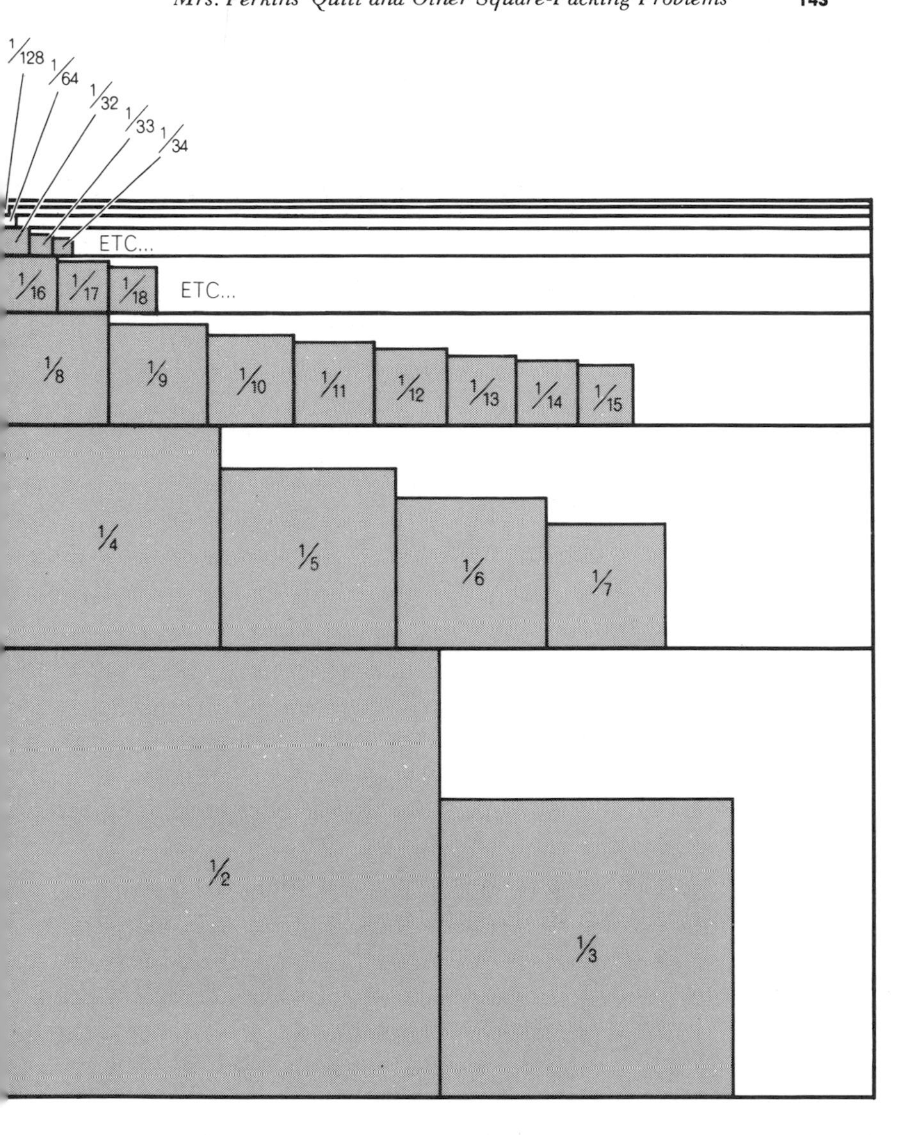

FIGURE 71

Packing an infinite set of squares into a unit square

In a 1967 paper Moser and his collaborator J. W. Moon, also at the University of Alberta, push the problem to its ultimate. They show that the infinite set of squares can be fitted into a square of side 5/6. (That no smaller square is possible is obvious, since the sum of the sides of the two largest squares is $1/2 + 1/3 = 5/6$.) A diagram of this tighter packing is given in the 1968 paper (see bibliography) by Moser and E. Meier. The surplus area is about 8 percent. Many other related results are given in the two papers on which Moser collaborated, including an elegant proof that any set of squares with a combined area of 1 can be packed without overlap into a square of area 2.

Among the many unsolved problems that concern the packing of squares into a larger square, one of the most infuriating is an unpublished problem proposed a few years ago by Richard B. Britton, of Carlisle, Mass. He had read Tutte's article in this department on squaring the square with unequal squares and wondered if it would be possible to divide a square into smaller squares with sides in serial order 1, 2, 3, 4, 5. . . . This would be possible, of course, only if the partial sum of the corresponding series of areas, $1 + 4 + 9 + 16 + 25 \ldots$, ever reaches a number that is itself a square. This does not happen until the first 24 square numbers have been added. The sum of $1^2 + 2^2 + 3^2 + \ldots + 24^2$ is 4,900, which is 70^2. Curiously, this never happens again.

The discovery of the uniqueness of 4,900 has an interesting history that involves a type of three-dimensional "figurate" number called a "pyramidal number." Pyramidal numbers are the cardinal numbers of sets of cannonballs that can be stacked into four-sided pyramids with no balls left over. Since each layer of such a pyramid is a square of balls, starting with one on top, then a layer of four, then a layer of nine, and so on, it is easy to see that a pyramidal number must be a partial sum of the series $1^2 + 2^2 + 3^2 + \ldots + n^2$. The formula for such a number can be written in this form:

$$\frac{n\,(n+1)\,(2n+1)}{6}.$$

An old puzzle asked for the smallest number of cannonballs that will form a four-sided pyramid and can also be rearranged flat on the ground to form a perfect square. In algebraic terms the problem asked for the smallest positive integral values of m and n that satisfy the Diophantine equation

$$\frac{n\,(n+1)\,(2n+1)}{6} = m^2.$$

The French mathematician Edouard Lucas, and later Dudeney, both conjectured that $n = 24$, $m = 70$ were the only positive integers that satisfied this equation, other than the trivial case of $n = 1$, $m = 1$. Put another way, 4,900 is the only number greater than 1 that is both square and pyramidal. It was not until 1918 that G. N. Watson (in *Messenger of Mathematics*, New Series, Vol. 48, pages 1–22) gave the first proof that this is indeed the case.

We know, therefore, that if a square checkerboard can be divided along lattice lines into squares with sides in the 1, 2, 3 . . . series, it must be the order-70 board. Although I know of no proof that this is impossible, the work of Tutte and others makes it extremely unlikely, and perhaps an impossibility proof would not be hard to find. The question now arises (and this is Britton's proposed problem): What is the largest area of the order-70 square that *can* be covered by squares taken from the set of 24 squares? Of course, not all the 24 squares can be used. It is assumed that no square overlaps the border of the order-70 square and that no two covering squares overlap each other.

The problem can be worked on by outlining an order-70 square on graph paper that has a matrix fine enough to make it possible, or one can paste down overlapping graph paper of larger mesh to make an order-70 square with unit squares of, say, a quarter inch on the side. The covering squares can be cut from thin cardboard. (It is not necessary to include the three smallest squares, which are too tiny to work with.) The best strategy is to place large squares first. At the end there are almost certain to be holes into which it will be obvious that the 1-, 2-, and 3-squares will fit.

Once you start pushing cardboard squares here and there over the order-70 square you are likely to get hooked on the problem. It has a peculiar fascination much like the challenge of packing as much as you can into a trunk or suitcase, but it has more mathematical precision. The exposed area is easily reduced to fewer than 200 unit squares. With ingenuity this can be chopped down to fewer than 150.

ADDENDUM

THE PROBLEMS in this chapter have obvious analogs with the packing of equilateral triangles inside equilateral triangles, and the packing of cubes inside cubes. Although I know of no published work in either field, the methods given in the papers by Conway and Trustrum can be applied to the triangular variant. Several readers wondered if Britton's problem had a cubical analog: Is there a cube which is the sum of the cubes of consecutive integers starting with 1? The answer is no. Indeed, 3,4,5, are known to be the only consecutive integers the sum of whose cubes is a cube. (See L. E. Dickson, *History of the Theory of Numbers*, Vol. II, pages 584–585.)

Two problems suggested by Britton's problem, both unsolved, were called to my attention by Solomon W. Golomb:

1. Is there a rectangle, other than 1×1, in which all squares of consecutive integers, side 1 through n, can be packed without overlap or excess space?

2. Is it possible to cover the plane by packing consecutive squares starting with 1?

If the answer to the first question is no, what is the smallest square, or rectangle, into which all squares of consecutive integers from 1 through n can be packed? Conway, Golomb, and

Robert Reid, of Lima, Peru, each spent some time on this question. For n through 17, the following table, supplied by Conway, lists minimum squares and their uncovered areas:

n	Side of square	Excess area
1	1	0
2	3	4
3	5	11
4	7	19
5	9	26
6	11	30
7	13	29
8	15	21
9	18	39
10	21	56
11	24	70
12	27	79
13	30	81
14	33	74
15	36	56
16	39	25
17	43	64

The minimum squares for $n = 18$ and for all higher n's are not known. Conway's upper and lower bounds show that the minimum square for $n = 18$ must be either of side 46 or 47. It is not hard to pack the 18 squares into the 47-square, which has an excess of 100. If the squares can be packed into the 46-square, the excess is only 7; so small that Conway doubts if it can be done.

ANSWERS

THE PROBLEM of cutting the order-13 square into 11 smaller squares is solved by the unique pattern shown in Figure 72. For readers interested in exploring higher-order squares, orders 14–

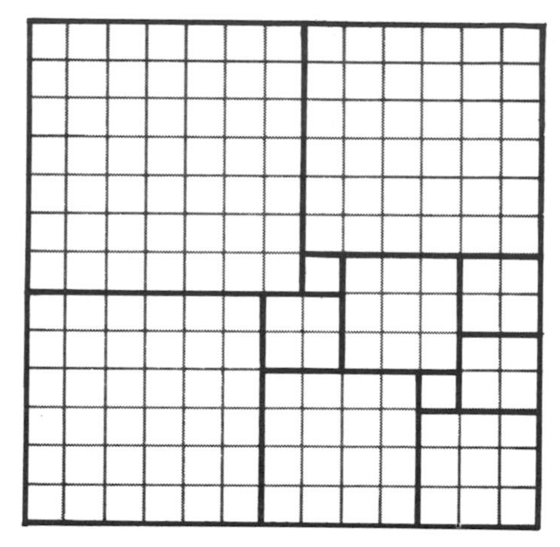

FIGURE 72
Solution to the order-13 square problem

17 are believed to have minimum patterns of 12 squares, orders 18–23 to have minimum patterns of 13 squares, orders 24–29 to have 14 squares, and orders 30–41 to have 15 squares, except that order 40 seems to need 16. Conway writes that the solution for order 41 is unusually hard to find. All orders through 100 need at most 19 squares.

About 250 readers sent solutions to the problem of covering as much as possible of the order-70 square with nonoverlapping squares from the set of 24 squares with sides 1, 2, 3 . . . 24. Almost all these solutions reduced the exposed area to fewer than 100 square units. Twenty-seven solutions were received that reduced the exposed area to 49 square units, or exactly 1 percent of the total area. All 27 patterns are identical (aside from rotations and reflections) in the placing of squares 11 through 24 (except for an interchange of squares 17 and 18), and in their omission of only the order-7 square. The first such solution received was from William Cutler. The solution given in Figure 73 came from Robert L. Patton.

In 1974 Edward M. Reingold, a computer scientist at the University of Illinois, Urbana, and his student, James Bitner, made an exhaustive computer search of the order-70 square for a perfect tiling with the set of 24 squares. The search proved that no such tiling is possible. Although their program is capable of

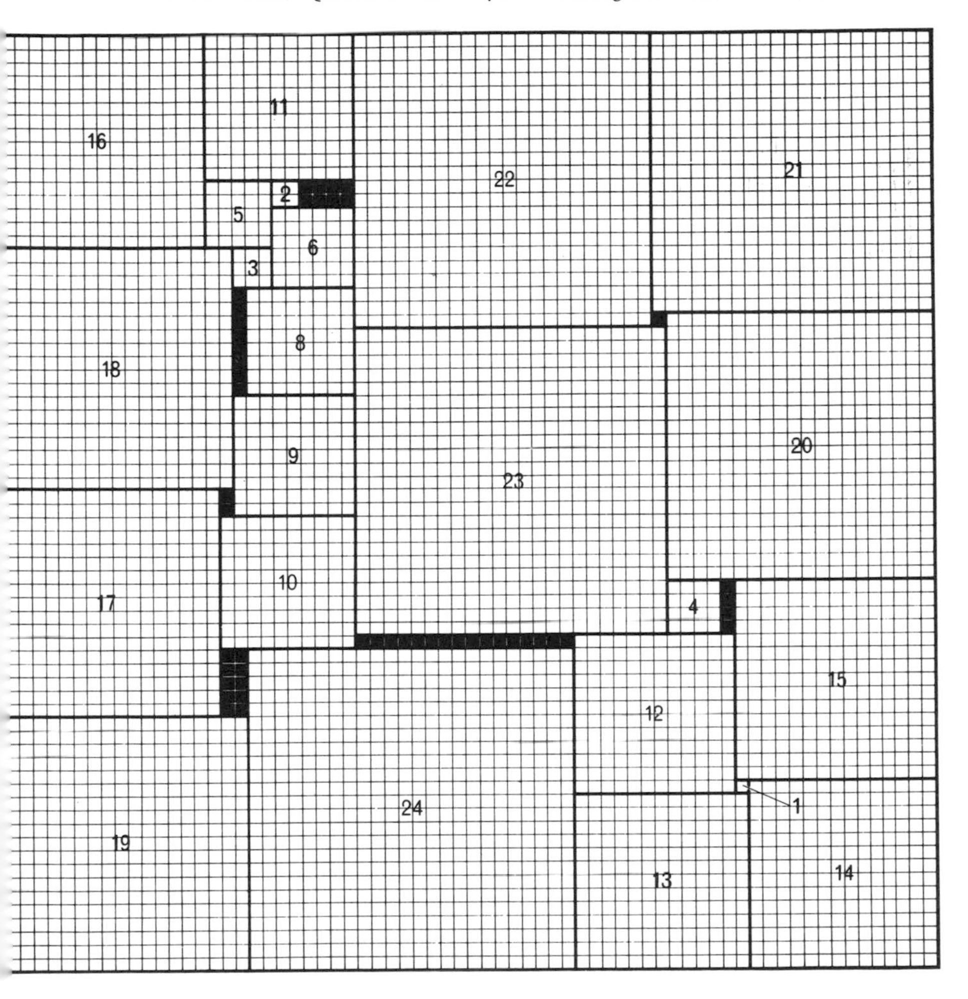

FIGURE 73
A solution to Britton's square-packing problem

finding the maximum area that can be tiled with squares from this set, it would take an infeasible amount of time to do so. It remains a conjecture, therefore, that the exposed area cannot be reduced below 49 square units.

CHAPTER 12

The Numerology of Dr. Fliess

> *At Aussee I know a wonderful wood full of ferns and mushrooms, where you shall reveal to me the secrets of the world of the lower animals and the world of children. I am agape as never before for what you have to say—and I hope that the world will not hear it before me, and that instead of a short article you will give us within a year a small book which will reveal organic secrets in periods of 28 and 23.*
>
> —SIGMUND FREUD, in a letter to Wilhelm Fliess, 1897

ONE OF the most extraordinary and absurd episodes in the history of numerological pseudoscience concerns the work of a Berlin surgeon named Wilhelm Fliess. Fliess was obsessed by the numbers 23 and 28. He convinced himself and others that behind all living phenomena and perhaps inorganic nature as well there are two fundamental cycles: a male cycle of 23 days and a female cycle of 28 days. By working with multiples of those two numbers—sometimes adding, sometimes subtracting—he was able to impose his number patterns on virtually everything. The work made a considerable stir in Germany during the early years of this century. Several disciples took up the system, elaborating and modifying it in books, pamphlets, and articles. In recent years the movement has taken root in the United States.

Although Fliess's numerology is of interest to recreational mathematicians and students of pathological science, it would probably be unremembered today were it not for one almost unbelievable fact: For a decade Fliess was Sigmund Freud's best friend and confidant. Roughly from 1890 to 1900, in the period of Freud's greatest creativity, which culminated with the publication of *The Interpretation of Dreams* in 1900, he and Fliess were linked in a strange, neurotic relationship that had—as Freud himself was well aware—strong homosexual undercurrents. The story was known, of course, to the early leaders of psychoanalysis, but few laymen had even heard of it until the publication in 1950 of a selection of 168 letters from Freud to Fliess, out of a total of 284 that Fliess had carefully preserved. (The letters were first published in German. An English translation entitled *The Origins of Psycho-Analysis* was issued by Basic Books in 1954.) Freud was staggered by the news that these letters had been preserved, and he begged the owner (the analyst Marie Bonaparte) not to permit their publication. In reply to her question about Fliess's side of the correspondence Freud said: "Whether I destroyed them [Fliess's letters] or cleverly hid them away I still do not know." It is assumed that he destroyed them. The full story of the Fliess-Freud friendship has been told by Ernest Jones in his biography of Freud.

When the two men first met in Vienna in 1877, Freud was thirty-one, relatively unknown, happily married, and with a modest practice in psychiatry. Fliess had a much more successful practice as a nose and throat surgeon in Berlin. He was two years younger than Freud, a bachelor (later he married a wealthy Viennese woman), handsome, vain, brilliant, witty, and well informed on medical and scientific topics.

Freud opened their correspondence with a flattering letter. Fliess responded with a gift, then Freud sent a photograph of himself that Fliess had requested. By 1892 they had dropped the formal *sie* (you) for the intimate *du* (thou). Freud wrote more often than Fliess and was in torment when Fliess was slow in answering. When his wife was expecting their fifth child,

Freud declared it would be named Wilhelm. Indeed, he would have named either of his two youngest children Wilhelm but, as Jones puts it, "fortunately they were both girls."

The foundations of Fliess's numerology were first revealed to the world in 1897 when he published his monograph *Die Beziehungen zwischen Nase und weibliche Geschlechtsorganen in ihrer biologischen Bedeutungen dargestellt* (*The Relations between the Nose and the Female Sex Organs from the Biological Aspects*). Every person, Fliess maintained, is really bisexual. The male component is keyed to the rhythmic cycle of 23 days, the female to a cycle of 28 days. (The female cycle must not be confused with the menstrual cycle, although the two are related in evolutionary origin.) In normal males the male cycle is dominant, the female cycle repressed. In normal females it is the other way around.

The two cycles are present in every living cell and consequently play their dialectic roles in all living things. Among animals and humans both cycles start at birth, the sex of the child being determined by the cycle that is transmitted first. The periods continue throughout life, manifesting themselves in the ups and downs of one's physical and mental vitality, and eventually determine the day of one's death. Moreover, both cycles are intimately connected with the mucous lining of the nose. Fliess thought he had found a relation between nasal irritations and all kinds of neurotic symptoms and sexual irregularities. He diagnosed these ills by inspecting the nose and treated them by applying cocaine to "genital spots" on the nose's interior. He reported cases in which miscarriages were produced by anesthetizing the nose, and he said that he could control painful menstruation by treating the nose. On two occasions he operated on Freud's nose. In a later book he argued that all left-handed people are dominated by the cycle of the opposite sex, and when Freud expressed doubts, he accused Freud of being left-handed without knowing it.

Fliess's theory of cycles was at first regarded by Freud as a major breakthrough in biology. He sent Fliess information on

23- and 28-day periods in his own life and the lives of those in his family, and he viewed the ups and downs of his health as fluctuations of the two periods. He believed a distinction he had found between neurasthenia and anxiety neurosis could be explained by the two cycles. In 1898 he severed editorial connections with a journal because it refused to retract a harsh review of one of Fliess's books.

There was a time when Freud suspected that sexual pleasure was a release of 23-cycle energy and sexual unpleasure a release of 28-cycle energy. For years he expected to die at the age of 51 because it was the sum of 23 and 28, and Fliess had told him this would be his most critical year. "Fifty-one is the age which seems to be a particularly dangerous one to men," Freud wrote in his book on dreams. "I have known colleagues who have died suddenly at that age, and amongst them one who, after long delays, had been appointed to a professorship only a few days before his death."

Freud's acceptance of Fliess's cycle theory was not, however, enthusiastic enough for Fliess. Abnormally sensitive to even the slightest criticism, he thought he detected in one of Freud's 1896 letters some faint suspicions about his system. This marked the beginning of the slow emergence of latent hostility on both sides. Freud's earlier attitude toward Fliess had been one of almost adolescent dependence on a mentor and father figure. Now he was developing theories of his own about the origins of neuroses and methods of treating them. Fliess would have little of this. He argued that Freud's imagined cures were no more than the fluctuations of mental illness, in obedience to the male and female rhythms. The two men were on an obvious collision course.

As one could have predicted from the earlier letters, it was Fliess who first began to pull away. The growing rift plunged Freud into a severe neurosis, from which he emerged only after painful years of self-analysis. The two men had been in the habit of meeting frequently in Vienna, Berlin, Rome, and elsewhere, for what Freud playfully called their "congresses." As late as 1900, when the rift was beyond repair, we find Freud writing: "There has never been a six months' period where I have longed more to be united with you and your family. . . . Your suggestion of a meeting at Easter greatly stirred me. . . . It is not merely my almost childlike yearning for the spring and for more beautiful scenery; that I would willingly sacrifice for the satisfaction of having you near me for three days. . . . We should talk reasonably and scientifically, and your beautiful and sure biological discoveries would awaken my deepest—though impersonal—envy."

Freud nevertheless turned down the invitation, and the two men did not meet until later that summer. It was their final meeting. Fliess later wrote that Freud had made a violent and unprovoked verbal attack on him. For the next two years Freud tried to heal the breach. He proposed that they collaborate on a book on bisexuality. He suggested that they meet again in 1902. Fliess turned down both suggestions. In 1904 Fliess published angry accusations that Freud had leaked some of his ideas to Hermann Swoboda, one of Freud's young patients, who in turn had published them as his own.

The final quarrel seems to have taken place in a dining room of the Park Hotel in Munich. On two later occasions, when Freud was in this room in connection with meetings of the analytical movement, he experienced a severe attack of anxiety. Jones recalls an occasion in 1912, when he and a group that included Freud and Jung were lunching in this same room. A break between Freud and Jung was brewing. When the two men got into a mild argument, Freud suddenly fainted. Jung carried him to a sofa. "How sweet it must be to die," Freud said as he was coming to. Later he confided to Jones the reason for his attack.

Fliess wrote many books and articles about his cycle theory, but his magnum opus was a 584-page volume, *Der Ablauf des Lebens: Grundlegung zur Exakten Biologie* (*The Rhythm of Life: Foundations of an Exact Biology*), published in Leipzig in 1906 (second edition, Vienna, 1923). The book is a masterpiece of Teutonic crackpottery. Fliess's basic formula can be written $23x + 28y$, where x and y are positive or negative integers. On almost every page Fliess fits this formula to natural phenomena, ranging from the cell to the solar system. The moon, for example, goes around the earth in about 28 days; a complete sunspot cycle is almost 23 years.

The book's appendix is filled with such tables as multiples of 365 (days in the year), multiples of 23, multiples of 28, multiples of 23^2, multiples of 28^2, multiples of 644 (which is 23×28). In boldface are certain important constants such as 12,167 [23×23^2], 24,334 [$2 \times 23 \times 23^2$], 36,501 [$3 \times 23 \times 23^2$], 21,952 [28×28^2], 43,904 [$2 \times 28 \times 28^2$], and so on. A table lists the numbers 1 through 28, each expressed as a difference between multiples of 28 and 23 [for example, $13 = (21 \times 28) - (25 \times 23)$]. Another table expresses numbers 1 through 51 [$23 + 28$] as sums and differences of multiples of 23 and 28 [for example, $1 = (\frac{1}{2} \times 28) + (2 \times 28) - (3 \times 23)$].

Freud admitted on many occasions that he was hopelessly deficient in all mathematical abilities. Fliess understood elementary arithmetic, but little more. He did not realize that if any two positive integers that have no common divisor are substituted for 23 and 28 in his basic formula, it is possible to express *any positive integer whatever*. Little wonder that the formula could be so readily fitted to natural phenomena! This is easily seen by working with 23 and 28 as an example. First determine what values of x and y can give the formula a value of 1. They are $x = 11$, $y = -9$:

$$(23 \times 11) + (28 \times -9) = 1.$$

It is now a simple matter to produce any desired positive integer by the following method:

$$[23 \times (11 \times 2)] + [28 \times (-9 \times 2)] = 2$$
$$[23 \times (11 \times 3)] + [28 \times (-9 \times 3)] = 3$$
$$[23 \times (11 \times 4)] + [28 \times (-9 \times 4)] = 4$$
$$\cdots$$

As Roland Sprague recently pointed out in a German puzzle book, even if negative values of x and y are excluded, it is still possible to express all positive integers greater than a certain integer. In the finite set of positive integers that *cannot* be expressed by this formula, asks Sprague, what is the largest number? In other words, what is the largest number that cannot be expressed by substituting nonnegative integers for x and y in the formula $23x + 28y$?

Freud eventually realized that Fliess's superficially surprising results were no more than numerological juggling. After Fliess's death in 1928 (note the obliging 28), a German physician, J. Aelby, published a book that constituted a thorough refutation of Fliess's absurdities. By then, however, the 23–28 cult was firmly established in Germany. Swoboda, who lived until 1963, was the cult's second most important figure. As a psychologist at the University of Vienna he devoted much time to investigating, defending, and writing about Fliess's cycle theory. In his own rival masterwork, the 576-page *Das Siebenjahr* (*The Year of Seven*), he reported on his studies of hundreds of family trees to prove that such events as heart attacks, deaths, and the onset of major ills tend to fall on certain critical days that can be computed on the basis of one's male and female cycles. He applied the cycle theory to dream analysis, an application that Freud criticizes in a 1911 footnote to his book on dreams. Swoboda also designed the first slide rule for determining critical days. Without the aid of such a device or the assistance of elaborate charts, calculations of critical days are tedious and tricky.

Incredible though it may seem, as late as the 1960's the Fliess system still had a small but devoted band of disciples in Germany and Switzerland. There were doctors in several Swiss hospitals who determined propitious days for surgery on the basis of Fliess's cycles. (This practice goes back to Fliess. In 1925, when Karl Abraham, one of the pioneers of analysis, had a gallbladder operation, he insisted that it take place on the favorable day calculated by Fliess.) To the male and female cycles modern Fliessians have added a third cycle called the intellectual cycle, which has a length of 33 days.

Two books on the Swiss system have been published here by Crown: *Biorhythm*, 1961, by Hans J. Wernli, and *Is This Your Day?*, 1964, by George Thommen. Thommen is the president of a firm that supplies calculators and charting kits with which to plot one's own cycles.

The three cycles start at birth and continue with absolute regularity throughout life, although their amplitudes decrease with old age. The male cycle governs such masculine traits as physical strength, confidence, aggressiveness, and endurance. The female cycle controls such feminine traits as feelings, intuition, creativity, love, cooperation, cheerfulness. The newly discovered intellectual cycle governs mental powers common to both sexes: intelligence, memory, concentration, quickness of mind.

On days when a cycle is above the horizontal zero line of the chart, the energy controlled by that cycle is being discharged. These are the days of highest vitality and efficiency. On days when the cycle is below the line, energy is being recharged. These are the days of reduced vitality. When your male cycle is high and your other cycles are low, you can perform physical tasks admirably but are low in sensitivity and mental alertness. If your female cycle is high and your male cycle low, it is a fine day, say, to visit an art museum but a day on which you are likely to tire quickly. The reader can easily guess the applications of other cycle patterns to other common events of life. I omit details about methods of predicting the sex of unborn children or computing the rhythmic "compatibility" between two individuals.

The most dangerous days are those on which a cycle, particularly the 23- or 28-day cycle, crosses the horizontal line. Those days when a cycle is making a transition from one phase to another are called "switch-point days." It is a pleasant fact that switch points for the 28-cycle always occur on the same day of the week for any given individual, since this cycle is exactly four weeks long. If your switch point for the 28-cycle is on

Tuesday, for instance, every other Tuesday will be your critical day for female energy throughout your entire life.

As one might expect, if the switch points of two cycles coincide, the day is "doubly critical," and it is "triply critical" if all three coincide. The Thommen and Wernli books contain many rhythmograms showing that the days on which various famous people died were days on which two or more cycles were at switch points. On two days on which Clark Gable had heart attacks, the second fatal, two cycles were at switch points. The Aga Khan died on a triply critical day. Arnold Palmer won the British Open Golf Tournament during a high period in July, 1962, and lost the Professional Golf Association Tourney during a triple low two weeks later. The boxer Benny (Kid) Paret died after a knockout in a match on a triply critical day. Clearly it behooves the Fliessian to prepare a chart of his future cycle patterns so that he can exercise especial care on critical days; since other factors come into play, however, no ironclad predictions can be made.

Because each cycle has an integral length in days, it follows that every person's rhythmogram will repeat its pattern after a certain interval of n days. This interval will be the same for everybody. For example, n days after every person's birth all three of his cycles will cross the zero line simultaneously on their upswing and his entire pattern will start over again. Two people whose ages are exactly n days apart will be running on perfectly synchronized cycle patterns. The reader should have no difficulty computing the value of n. It is an important constant in the Swiss Fliessian system.

ADDENDUM

GEORGE S. THOMMEN, president of Biorhythm Computers, Inc., 298 Fifth Avenue, New York, is still going strong, appearing occasionally on radio and television talk shows to promote his products. James Randi, the magician, was moderator of an all-night radio talk show in the mid-sixties. Thommen was twice his guest. After one of the shows, Randi tells me, a lady in New Jersey sent him her birth date and asked for a biorhythm chart covering the next two years of her life. After sending her an actual chart, but based on a *different* birth date, Randi received an effusive letter from the lady saying that the chart exactly matched all her critical up and down days. Randi wrote back, apologized for having made a mistake on her birth date, and enclosed a "correct" chart, actually as wrongly dated as the first one. He soon received a letter telling him that the new chart was even *more* accurate than the first one.

Speaking in March, 1966, at the 36th annual convention of the Greater New York Safety Council, Thommen reported that biorhythm research projects were underway at the University of Nebraska and the University of Minnesota, and that Dr. Tatai, medical chief of Tokyo's public health department, had published a book, *Biorhythm and Human Life*, using the Thommen system. When a Boeing 727 jetliner crashed in Tokyo in February, 1966, Dr. Tatai quickly drew up the pilot's chart, Thommen said, and found that the crash occurred on one of the pilot's low days.

Biorhythm seems to have been more favorably received in Japan than in the United States. According to *Time*, January 10, 1972, page 48, the Ohmi Railway Co., in Japan, computed the biorhythms of each of its 500 bus drivers. Whenever a driver was scheduled for a "bad" day, he was given a notice to be extra careful. The Ohmi company reported a fifty percent drop in accidents.

Fate magazine, February, 1975, pages 109–110, reported on a conference on "Biorhythm, Healing and Kirlian Photography," held in Evanston, Ill., October, 1974. Michael Zaeske, who sponsored the conference, revealed that the traditional biorhythm curves are actually "first derivatives" of the true curves, and that all the traditional charts are "in error by several days." Guests at the meeting also heard evidence from California that a fourth cycle exists, and that all four cycles "may be related to Jung's four personality types."

Science News, January 18, 1975, page 45, carried a large ad by Edmund Scientific Company for their newly introduced Biorhythm Kit ($11.50), containing the precision-made Dialgraf Calculator. The ad also offered an "accurate computerized, personalized" biorhythm chart report for 12 months to any reader who sent his birthdate and $15.95. One wonders if Edmund is using the traditional charts (possibly off three days) or Zaeske's refined procedures.

ANSWERS

THE LARGEST positive integer that cannot be expressed as a sum of multiples of two nonnegative integers a and b that are relatively prime is equal to $ab - a - b$. In the case at hand: $(23 \times 28) - 23 - 28 = 593$. For a proof of the formula, see the solution to Problem 26 in Roland Sprague's *Recreation in Mathematics* (London: Blackie, 1963).

The second problem was to determine when a person's biological chart, as worked out by the Swiss school based on the work of Wilhelm Fliess, will finish a complete cycle and start repeating the same pattern. The three superimposed cycles have periods of 23, 28, and 33 days. These numbers are prime to each other (have no common divisor) and so the combined pattern will not repeat until after a lapse of $23 \times 28 \times 33 = 21{,}252$ days, or a little more than 58 years. Since Fliess's system did not include the 33-day cycle, his cycle patterns repeat after a lapse of $23 \times 28 = 644$ days. Swiss Fliessians call this the "biorhythmic year." It is important in computing the "biorhythmic compatibility" between two individuals, since any two persons born 644 days apart are synchronized with respect to their two most important cycles.

CHAPTER 13

Random Numbers

And the earth was without form, and void; and darkness was upon the face of the deep.

GENESIS 1:2

A BOOK called *A Million Random Digits with 100,000 Normal Deviates* was prepared by the Rand Corporation and published in 1955 by the Free Press, now a division of Macmillan. A specimen page consists of nothing but repetitions of the 10 digits, 0 through 9. They are printed on the page in very orderly fashion, in groups of five, but the *sequence* of digits is as disheveled as Rand mathematicians could make it.

"The production of such a book is entirely of the twentieth century," writes the physicist Alfred M. Bork in an article titled "Randomness and the Twentieth Century" that appeared in *The Antioch Review*, Spring, 1967. "It could not have been produced in any other era. I do not mean to stress that the mechanism for doing it was not available, although that is also true. What is of more interest is that before the twentieth century no one would even have thought of the *possibility* of producing a book like this; no one would have seen any use for it. A rational nineteenth-century man would have thought it the height of folly. . . ."

It is Bork's thesis that preoccupation with randomness has permeated 20th-century culture. This preoccupation has several 19th-century scientific sources—chiefly thermodynamics, in which entropy is a measure of disorder, and the theory of evolution, in which natural selection imposes orderly development on random mutations. Early in this century randomness became the bedrock of quantum mechanics, an irreducible chance element in the microstructure of the world. Eventually it may turn out that behind this apparent haphazardry there are nonrandom laws (as Einstein believed; he found displeasing the notion, as he once expressed it, of God's playing dice), but at present no one knows what those laws are and, if they *are* ever found, quantum theory will have to be replaced by a radically different theory. Bork sees an influence of these scientific ideas on the random art of abstract expressionism, in the random music of such composers as John Cage, in the random wordplay of such books as *Finnegans Wake,* and in William Burroughs' technique of cutting up the pages of a novel, shuffling the pieces and then printing them in a random order.

Perhaps, too, some artists find in haphazardry a relief from the excessive orderliness of modern technology. Lord Dunsany has a beautiful description (in his *Tales of Three Hemispheres*) of a visit to New York City during which he becomes oppressed by the monotonous right-angled regularity of the city's streets and the dull orthogonal arrangements of the windows of its tall buildings. Slowly dusk comes and those windows begin to glow in irregular patterns. "Surely if modern man with his clever schemes held any sway here still he would have turned one switch and lit them all together; but we are back with the older man of whom far songs tell, he whose spirit is kin to strange romances and mountains. One by one the windows shine from the precipices; some twinkle, some are dark; man's orderly schemes have gone, and we are amongst vast heights lit by inscrutable beacons. . . . Here in New York a poet met a welcome."

A random pattern of lighted windows is a geometric counterpart to a sequence of random digits. What exactly is such a sequence? It is curiously hard to say. One ordinarily calls a finite series of digits random if, given all but one digit in the series, there is no rule by which the missing digit can be guessed with a probability of better than 1/10. But this is a subjective definition, based on one's ignorance of possible underlying patterns.

Is there any objective, mathematical way to define a completely disordered series? Apparently there is not. The best one can do is to specify certain tests for types of randomness and call a series random to the degree that it passes them. For example, one can insist that a series meet the following formal criteria: each digit, or "atomic unit," appears with a frequency of 1/10, each permutation of digits taken two at a time appears, among all couplets in the series, with a frequency of 1/100, each permutation of digits three at a time appears with a frequency of 1/1,000, and so on for all higher "molecular units." The couplets, triplets, quartets and so forth are not confined to adjacent digits. A test for the randomness of, say, triplets could pick three digits separated by any specified intervals.

An infinite decimal fraction between 0 and 1 that satisfies such a test (of course in practice there can only be minute partial testing; a complete test would require an infinite time) is said to be a "normal number." Clearly the decimal expression for any rational fraction is not normal; it endlessly repeats a sequence, such as 1/3, which keeps repeating 3, or 1/97, which repeats a sequence of 96 digits. But the decimal expressions of irrational numbers such as the square root of 2, and such famous transcendental irrationals as pi and *e*, are believed to be "normal." (A transcendental number is an irrational number that is not the root of an algebraic equation but must be expressed as the limit of an infinite converging series.) At least they have so far passed all tests for normality.

It has been proved that among the infinite number of decimal expressions for the real fractions between 0 and 1 infinitely more are normal than not. Pick a real number at random and the probability is zero (here *probability* is used in a special sense) that you will pick one that is not normal. Can we say that the sequence of digits in a normal decimal fraction is random? Sometimes. Put a decimal point in front of the first of Rand's one million random digits and you have the beginning of an infinite number of normal decimal fractions. On the other hand, the decimal expression of pi, calculated in 1974 to a million places, has satisfied all tests for normality and yet cannot be called a random sequence because it can be constructed as the limit of simple formulas. Every next digit in pi is predictable with certainty. Because it is pi, therefore, it is highly ordered, although aside from being pi it seems to have no discernible regularities.

It may surprise some readers to know that it is easy to construct infinite decimal fractions that are irrational but have obvious patterns. A simple example using only 0 and 1 is

$$.101001000100001000001\ldots,$$

in which the first 1 is followed by one 0, the second 1 by two 0's, the third by three 0's and so on. Since there is no repeating sequence the number is irrational. This is one of innumerable ways that irrational numbers with simple regularities can be written.

Many patterned numbers of this type can even be proved transcendental. Indeed, the first proof that transcendental numbers existed was given by a 19th-century French mathematician, Joseph Liouville, who found an infinite set of such numbers—now called Liouville numbers—that he proved to be transcendental. An interesting example of a number that has been proved both normal and transcendental, yet is so simply patterned that a child can write it, is obtained merely by putting down the counting numbers in order:

$$.12345678910111213141516171819.\ldots$$

Most mathematicians now agree that an absolutely disordered series of digits is a logically contradictory concept. A series can no more be patternless than an arrangement of stars in the sky can be. The reason in both cases is that as a series of digits or an arrangement of points comes closer and closer to satisfying all tests for randomness it begins to exhibit a very rare and unusual type of statistical regularity that in some cases even permits the prediction of missing portions.

To take a simple instance, suppose you are asked to fill 10 spaces in a row with digits and do it in a completely disordered way. If you duplicate one or more digits, the series will be ordered in the sense that it shows a bias toward those digits. On the other hand, if it is completely free from *that* kind of bias, it will contain each of the 10 digits. Such a series will satisfy absolutely the criterion that there be no bias for any one digit, but for this a price has been paid: the series is now so strongly "patterned" that, given any nine digits, the missing digit can be guessed with a probability of 1. Similar contradictions turn up in connection with any random series. If it gets too random, a "pattern of disorder," so to speak, appears.

We thus face a curious paradox. The closer we get to an absolutely patternless series, the closer we get to a type of pattern so rare that if we came on such a series we would suspect it had been carefully constructed by a mathematician rather than produced by a random procedure. We can speak of a series of digits as being disorderly only in a relative sense, that is, disordered with respect to tests of certain kinds but not to tests of other kinds.

The whole matter is spattered with profound difficulties. G. Spencer Brown, in his book *Probability and Scientific Inference* (1957), pointed out some of these paradoxes and showed how easily one could take printed tables of random numbers and find various kinds of order if one looked hard enough. Many of the published results of extrasensory-perception testing, Brown argued convincingly, are examples of patterns that are inevitable in any long series of random results. When such patterns fail to turn up, ESP proponents are unlikely to publish the results, not having found evidence for ESP; when they do find such patterns, they publish. Perhaps if the total picture could be surveyed, the published patterns would be less surprising.

As a puzzle interlude at this point the reader is invited to study the following apparently patternless arrangement of the 10 digits:

7480631952.

By what rule are those digits ordered? Hint: The arrangement is cyclic. Think of the head and tail of the sequence as being joined to make a circle.

It is possible, of course, that such a sequence, or any other strongly ordered sequence of 10 digits, might accidentally turn up somewhere in Rand's one million digits. If a series of random digits is long enough, such surprising patterns are certain to be found in it. Some philosophers have argued that the universe is like this: an accidental segment of order in a vast, infinite sea of chaos. Jorge Luis Borges has given this a classic metaphorical expression in his famous short story "The Library of Babel." Existence is a nonsense collection of all possible combinations of whatever the basic micro building blocks are. The little spot of accidental order that is our universe is like the sequence 12345-6789 in an infinite series of random digits.

At this point an ancient philosophical controversy arises. Why is a pattern such as 123456789 in a table of random digits "surprising"? It is neither more nor less probable than any other permutation of nine digits. Certain pragmatists and subjectivists have argued that the concept of "pattern" in any arrangement of parts cannot be defined except with reference to human experience. The only reason we say the first one million decimals of pi are ordered and the Rand digits are not is that pi is a useful constant for man.

"Order and disorder," wrote William James in his *Varieties of Religious Experience* (he later changed his mind), "are purely human inventions. . . . If I should throw down a thousand beans at random upon a table, I could doubtless, by eliminating a sufficient number of them, leave the rest in almost any geometrical pattern you might propose to me, and you might then say that that pattern was the thing prefigured beforehand, and that the other beans were mere irrelevance and packing material. Our dealings with Nature are just like this. She is a vast *plenum* in which our attention draws capricious lines in innumerable directions. We count and name whatever lies upon the special lines we trace, whilst the other things and the untraced lines are neither named nor counted."

To this argument the realist replies that it is just the other way around. Instead of our brain's imposing its patterns on nature, the brain is at birth merely an intricate net of random connections. It acquires its ability to "see" patterns only after years of experience during which the patterned external world imposes its order on the brain's *tabula rasa*. It is true, of course, that one is surprised by a sequence of 123456789 in a series of random digits because such a sequence is defined by human mathematicians and used in counting, but there is a sense in which such sequences correspond to the structure of the outside world. Starting at a given point of time in the distant past, before life existed on the earth, the moon circled the earth once, then twice, then three times and so on, even though no human observers were there to count. At any rate, ordinary language as well as the language of science enables one to make such statements, and my own view is that only confusion results when one tries to adopt a language in which one cannot speak of the universe as patterned apart from human observation.

Let us get back to a less metaphysical question. How are tables of random digits produced? It is no good just scribbling digits on paper as fast as they pop into your head; humans are incapable of producing them at random. Too many unconscious biases creep into such a series. You might suppose you could take a table of, say, logarithms or the populations of American cities in alphabetical order, and copy down the first digits of those numbers. But it was discovered about 20 years ago that the first digits of *any* table of random numbers show a marked bias: the lower the digit, the higher its frequency in the table! Warren Weaver has an excellent exposition of this astonishing fact in his paperback *Lady Luck: The Theory of Probability*, pages 270–277.

One way to get a series of random digits is by using a physical process involving so many variables that the next digit can never be predicted with a probability higher than $1/n$, where n is the base of the number system used. Flipping a penny generates a random series of binary digits. A perfect die randomizes six symbols. A 10-position spinner, or an icosahedral die with each digit appearing twice on its 20 faces, will randomize the 10 digits. A regular dodecahedron is an excellent randomizer for the 12 digits of a base-12 number system. One can even get down to the chance level of quantum mechanics and base a randomizer on the timing of the clicks of a Geiger counter as it records radioactive decay.

There are many other ways. In 1927 L. H. C. Tippett published 41,600 random numbers by taking the middle digits of the areas of parishes in England. In 1939 a table of 100,000 random digits was produced by M. G. Kendall and B. Babington Smith. They used a roulette wheel with a rim marked off into 10 parts. While the wheel spun rapidly they illuminated it by hand with flashes of light and recorded the digit at a certain spot. In 1949 the U.S. Interstate Commerce Commission extracted 105,000 random digits from freight waybills. Rand's one million digits were obtained by electronic-pulse methods that produced random binary digits, which were then converted into decimal digits. To remove slight biases found by intensive testing, the one million digits were further randomized by adding all pairs and retaining only the last digit.

When a computer needs random numbers for solving a problem, it is less costly to let the machine generate its own series rather than take up valuable memory storage space by feeding it a published table. There are hundreds of ways in which computers can generate what are called "pseudo-random" digits. Calculating an irrational number such as pi or the square root of 3 is a poor method because it would take too long, and would require too much valuable storage space. An early procedure, proposed by John von Neumann, was the "middle of the square" method. The computer starts with a number of n digits, squares it, takes the middle n or $n + 1$ digits of the result, squares that, takes the middle digits again, and in this way continues to generate groups of n digits. This method is no longer used because it produces sequences that are too short, and because it was found to introduce too many biases. Birger Jansson, in his book *Random Number Generators* (1966), calls attention to some amusing anomalies that turned up. If you start with the number 3792 and square it, you get 14379264, so that your "random" series proves to be 3792 3792 3792. . . . The same thing happens if you start with such six-digit numbers as 495475 and 971582. Modern techniques for generating pseudorandom numbers are much superior and fantastically rapid, and they vary from one computer center to another.

A final word about the increasingly important uses for random numbers. They are indispensable in the designing of experiments in agriculture, medicine, and other fields where certain variables must be randomized to eliminate bias. They are used in game and conflict situations in which the best play is obtained by a random mixing of strategies. Above all they are essential for simulating and solving a variety of difficult problems involving complex physical processes in which random events play a major role. As Robert R. Coveyou, a mathematician at the Oak Ridge National Laboratory, put it recently: "The generation of random numbers is too important to be left to chance."

ADDENDUM

ONE OF the most promising of recent attempts to give a precise definition of a "random" or "patternless" sequence of digits is a proposal that was made independently by A. N. Kolmogorov in Russia in 1965 and by G. J. Chaitin, of IBM, in 1966. In essence, the "randomness" of a chain of digits is defined by the length of the shortest program that will tell a Turing machine (idealized digital computer) how to write the given chain.

In terms of information theory, we can put it this way. If an output sequence of digits consists of k bits, it can be obtained by an input of *k* or fewer bits. The greater the "order" of the digits, the smaller the required program. If the output is a highly ordered chain, such as 12121212, it can be obtained by a much shorter program than almost any set of eight digits picked at random from the Rand table of random numbers. Number sequences which require programs of maximum length are sequences in which there is no "pattern," therefore no way to shorten the required program. The disorder of a chain is measured by the length of the shortest program required to generate it. No finite chain of digits can be absolutely patternless, but we can think of absolute disorder as a limit concept. A very long string of digits generated by a good randomizer is almost certain to be extremely close to the disorder limit.

Readers interested in this approach to randomness can consult a nontechnical exposition by Chaitin, "Information-Theoretic Computational Complexity," in the *IEEE Transactions on Information Theory*, Vol. IT-20, January, 1974, pages 10–15; Per Martin-Löf, "The Definition of Random Sequences," *Information and Control*, Vol. 9, 1966, pages 602–619; and Terry Fine's book, *Theories of Probability: An Examination of Foundations*, Academic Press, 1973.

In all discussions of random numbers it is important to keep in mind that the word "random" is sometimes used to describe any sequence obtained by a randomizer, and sometimes to describe the absence of a pattern in a given sequence, and that these two uses of the word are not the same. For example, if you flip a penny six times and happen to get *HHHHHH*, the series is random in the first sense (*HHHHHH* is just as likely to occur as any other combination) but not in the second sense. The Turing-machine approach to randomness is a way of defining what is meant by saying that a sequence has maximum disorder or patternlessness. It is no help in generating random numbers.

Ean Wood, a London reader, disagreed with my assertion that *Finnegans Wake* contains random wordplay. "Each word," Wood declared, "is the result of careful Joyce."

ANSWERS

THE PROBLEM was to find the rule for the cyclic ordering of the 10 digits in 7480631952. Start at the left and spell *zero*, counting one digit for each letter. The spelling ends on 0. Cross it out. Continue with *one*. Cross out 1. Proceed in this manner, spelling the digits in order from 0 through 9, counting only the digits that have not already been crossed out. The series is circular; if a count is uncompleted at the end of the line of digits, go back to the beginning. The arrangement makes it possible for all 10 digits to be spelled in numerical order.

Sets of playing cards can be similarly arranged so that each card can be spelled by moving cards singly from the top to the bottom of a packet, discarding the card that turns up at the end of each spelling. It is easy to construct such arrangements simply by time-reversing the procedure, taking the cards one at a time from a pile, in reverse order, and forming the packet in your hands by moving the cards singly from the bottom to the top of the packet for each letter. To test this simple procedure the reader may enjoy arranging an entire deck so that every card can be spelled, starting with the Ace of spades and then proceeding with the two of spades and so on, the spades being followed by the other suits in a predetermined order, and ending with, say, the King of diamonds and the Joker.

Many readers correctly pointed out that the spelling procedure is only *one* rule for forming the sequence 7480631952. Any given finite string of digits can be generated by an infinity of rules, although the longer the sequence, the more complicated the rules usually become. If we take 7480631952 as an integer, there is an infinity of equations for which this integer is the solution, though unlikely that any of them could be called "simple." Put a decimal point in front of the number and we have the beginning of an infinity of decimal fractions, each one of which is the solution to an infinity of equations. Indeed, .7480631952 . . . is extremely close to the square root of 10 minus the square root of 2.

Mrs. Myron Milbouer, of Wilmington, Del. found a surprisingly simple way to generate the sequence 7480631952. Write the digits in triangular formation:

```
   1
  2 3
 4 5 6
7 8 9 0
```

The triangle has four diagonals that run from *NW* to *SE*. Starting at the left, take the first two diagonals downward and you get 7 (the first diagonal), then 4, 8. To this, add the next two diagonals, taking them in reverse fashion, from right to left and *upward*, and you get 0,6,3,1 and 9,5,2. A truly astonishing coincidence!

CHAPTER 14

The Rising Hourglass and Other Physics Puzzles

THE PROBLEMS that follow are not so much mathematical as they are problems combining logical reasoning with some knowledge (mostly elementary) of physical laws. They could be called physics puzzles. In a few cases the questions are worded so as to misdirect the reader, but there are no joke solutions that depend on language quibbles. I want to thank David B. Eisendrath, Jr., John B. Hart, Jerome E. Salny, Dave Fultz, and Derek Verner for problems 1, 12, 15, 18, and 23 respectively.

1. TWO HUNDRED PIGEONS

AN OLD story concerns a truck driver who stopped his panel truck just short of a small, shaky-looking bridge, got out and began beating his palms against the sides of the large compartment that formed the back of the truck. A farmer standing at the side of the road asked him why he was doing this.

"I'm carrying 200 pigeons in this truck," explained the driver. "That's quite a lot of weight. My pounding will frighten the birds and they'll start flying around inside. That will lighten the load considerably. I don't like the looks of this bridge. I want to keep those pigeons in the air until I get across."

Assuming that the truck's compartment is airtight, can anything be said for the driver's line of reasoning?

2. THE RISING HOURGLASS

AN UNUSUAL toy is on sale in Paris shops: a glass cylinder, filled with water, at the top of which an hourglass floats [*see Figure 74*]. If the cylinder is inverted, as in the right-hand

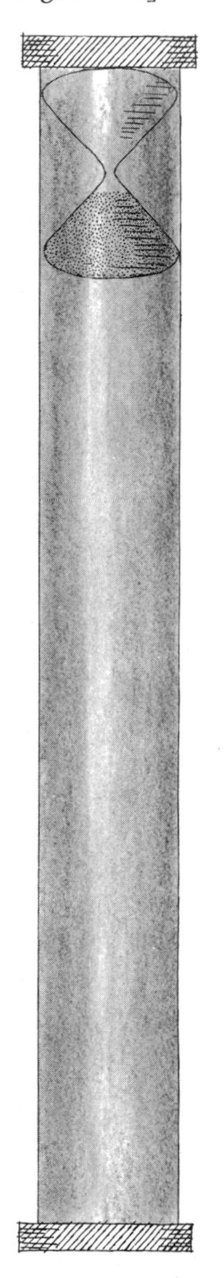

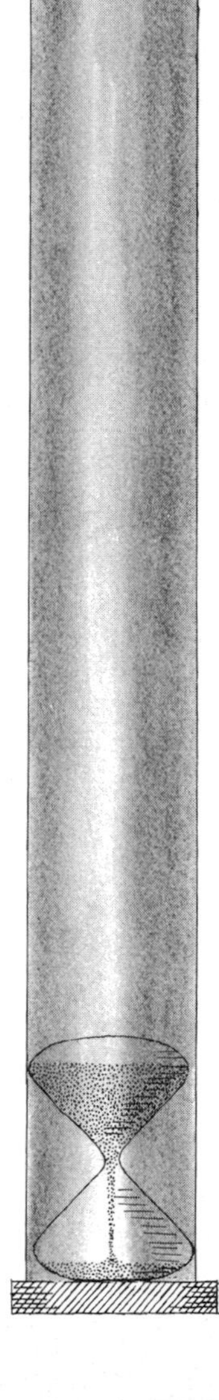

FIGURE 74
The hourglass paradox

drawing, a curious thing happens. The hourglass remains at the bottom of the cylinder until a certain quantity of sand has flowed into its lower compartment, then it rises slowly to the top. It seems impossible that a transfer of sand from top to bottom of the hourglass would have any effect on its overall buoyancy. Can you guess the simple *modus operandi?*

3. IRON TORUS

A PIECE of solid iron in the form of a doughnut is heated. Will the diameter of its hole get larger or smaller?

4. SUSPENDED HORSESHOE

FROM A sheet of thin pasteboard cut a horseshoe shape that is a trifle longer than a toothpick [*see Figure* 75]. Lean the pick and horseshoe together on a tablecloth as shown. The problem is to lift both the horseshoe and the toothpick by means of a second toothpick held in one hand. You may not touch the horseshoe or the toothpick on which it leans with anything except the toothpick in your hand. You are not, of course, permitted to break the toothpick and use the pieces like miniature chopsticks. Both objects must be lifted together and held suspended above the table. How is it done?

FIGURE 75
Horseshoe and toothpick

5. CENTER THE CORK

FILL A glass with water and drop a small cork on the surface. It will float to one side, touching the glass. How can you make it float permanently in the center, not touching the glass? The glass must contain nothing but water and the cork.

6. OIL AND VINEGAR

SOME FRIENDS were picnicking. "Did you bring the oil and vinegar for the salad?" Mrs. Smith asked her husband.

"I did indeed," replied Mr. Smith. "And to save myself the trouble of carrying two bottles I put the oil and vinegar in the same bottle."

"That was stupid," snorted Mrs. Smith. "I like a lot of oil and very little vinegar, but Henrietta likes a lot of vinegar, and——"

"Not stupid at all, my dear," interrupted Mr. Smith. He proceeded to pour, from the single bottle, exactly the right proportions of oil and vinegar that each person wanted. How did he manage it?

7. CARROLL'S CARRIAGE

IN CHAPTER 7 of Lewis Carroll's *Sylvie and Bruno Concluded*, the German Professor explains how people in his country need not go to sea to enjoy the sensations of pitching and rolling. This is accomplished, he says, by putting oval wheels on their carriages. An Earl, who is listening, says that he can see how oval wheels could make a carriage pitch backward and forward, but how could they make the carriage roll too?

"They do not *match*, my Lord," the Professor replies. "The *end* of one wheel answers to the *side* of the opposite wheel. So first one side of the carriage rises, then the other. And it pitches all the while. Ah, you must be a good sailor, to drive in our boat-carriages!"

Is it possible to arrange four oval wheels on a carriage so that it would actually pitch and roll as described?

Figure 76 reproduces a poem written and illustrated by Gelett Burgess (from *The Purple Cow and Other Nonsense*, Dover, 1961). Perhaps it was inspired by Lewis Carroll's carriage.

FIGURE 76

8. MAGNET TESTING

You are locked in a room that contains no metal of any sort (not even on your person) except for two identical iron bars. One is a bar magnet, the other is not magnetized. You can tell which is the magnet by suspending each by a thread tied around its center and observing which bar tends to point north. Is there a simpler way?

9. MELTING ICE CUBE

A cube of ice floats in a beaker of water, the entire system at 0 degrees centigrade. Just enough heat is supplied to melt the cube without altering the system's temperature. Does the water level in the beaker rise, fall, or stay the same?

10. STEALING BELL ROPES

In the tower of a church two bell ropes pass through small holes a foot apart in a high ceiling and hang down to the floor of a room. A skilled acrobat, carrying a knife and bent on stealing as much of the two ropes as possible, finds that the stairway leading above the ceiling is barred by a locked door. There are no ladders or other objects on which he can stand, and so he must accomplish his theft by climbing the ropes hand over hand and cutting them at points as high as possible. The ceiling is so high, however, that a fall from even one-third the height could be fatal. By what procedure can he obtain a maximum amount of rope?

11. MOVING SHADOW

A man walking at night along a sidewalk at a constant speed passes a street light. As his shadow lengthens, does the top of the shadow move faster or slower or at the same rate as it did when it was shorter?

12. THE COILED HOSE

A GARDEN hose is coiled around a reel about a foot in diameter placed on a bench as shown in Figure 77. One end of the hose hangs down into a bucket and the other end is unwound so that it can be held several feet above the reel. The hose is empty and there are no kinks or obstructions in it. If water is now poured into the upper end by means of a funnel, one would expect that continued pouring would cause water to run out at the lower end. Instead, as water is poured into the funnel it rises in the raised end of the hose until it overflows the funnel; no water ever emerges from the other end. Explain.

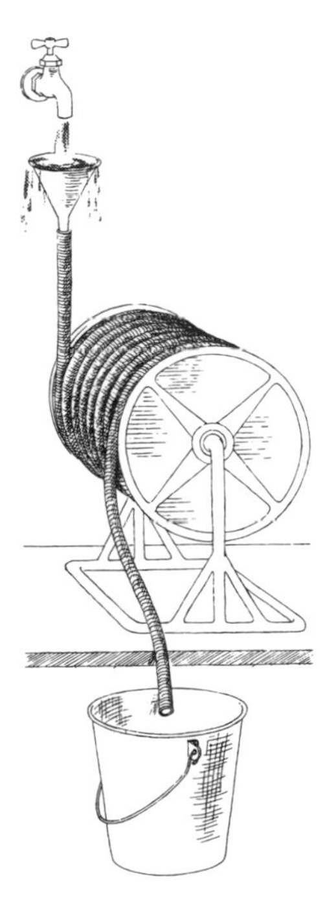

FIGURE 77
The hose paradox

13. EGG IN BOTTLE

YOU CANNOT push a peeled hardboiled egg through the neck of a glass milk bottle because the air trapped in the bottle keeps it from going through. If, however, you drop a piece of burning paper or a couple of burning matches into the bottle before you stand the egg upright at the opening, the burning will heat and expand the air. When the air cools it contracts, forming a partial vacuum which draws the egg inside. After this has happened a second problem presents itself. Without breaking the bottle or damaging the egg, how can you get it out again?

14. BATHTUB BOAT

A SMALL boy is sailing a plastic boat in the bathtub. It is loaded with nuts and bolts. If he dumps all this cargo into the water, allowing the boat to float empty, will the water level in the tub rise or fall?

15. BALLOON IN CAR

A FAMILY is out for a drive on a cold afternoon, with all vents and windows of the car closed. A child in the back seat is holding the lower end of a string attached to a helium-filled balloon. The balloon floats in the air, just below the car's roof. When the car accelerates forward, does the balloon stay where it is, move backward, or move forward? How does it behave when the car rounds a curve?

16. HOLLOW MOON

IT HAS been suggested that in the far future it may be possible to hollow out the interior of a large asteroid or moon and use it as a mammoth space station. Assume that such a hollowed asteroid is a perfect, nonrotating sphere with a shell of constant thickness. Would an object inside, near the shell, be pulled by the shell's gravity field toward the shell or toward the center of of the asteroid, or would it float permanently at the same location?

17. LUNAR BIRD

A BIRD has a small, lightweight oxygen tank attached to its back so that it can breathe on the moon. Will the bird's flying speed on the moon, where the pull of gravity is less than on the earth, be faster, slower, or the same as its speed on the earth? Assume that the bird carries the same equipment in both instances.

18. THE COMPTON TUBE

A LITTLE-KNOWN invention of the physicist Arthur Holly Compton is shown in Figure 78. A glass torus several feet across is completely filled with a liquid in which small particles are suspended. The tube is allowed to rest until there is no movement of its liquid, then it is quickly flipped upside down by a

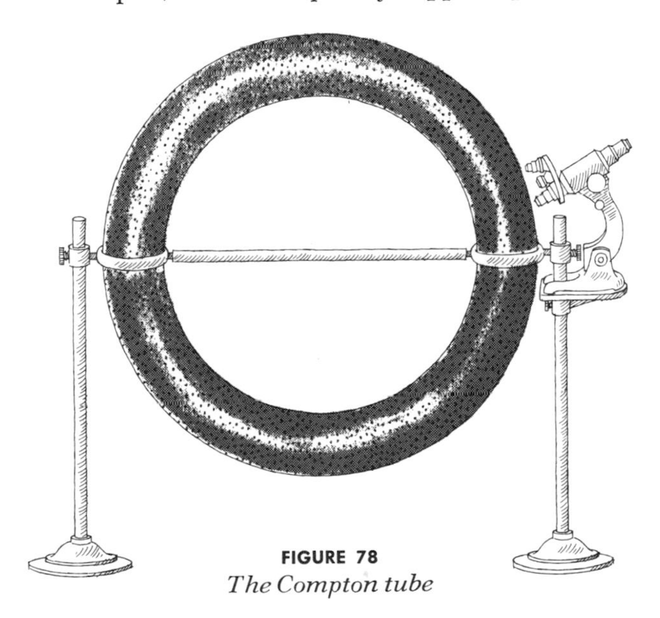

FIGURE 78
The Compton tube

180-degree rotation about the horizontal axis. By viewing the suspended particles through a microscope one can determine whether or not the liquid is now flowing around the torus.

Assume that the tube is oriented so that its vertical plane extends east and west. As the earth rotates counterclockwise (as seen looking down at the North Pole), the top of the tube moves faster than the bottom because it travels a circular path of larger circumference. Flipping the tube brings this faster-moving liquid to the bottom and the slower-moving liquid to the top, setting up a very weak clockwise circulation. The strength of this circulation diminishes as the plane of the tube deviates from east-west, reaching zero when the plane is north-south. One can, therefore, prove that the earth rotates and determine the direction of rotation simply by flipping the tube in different orientations until a maximum circulating speed is produced.

In actual practice, the viscosity of the liquid causes the circulation to decay after 20 seconds or so. Assuming that there is no friction whatever in the tube, and that the flip is made at the equator with an east-west orientation, how long will it take a particle in the liquid to complete one rotation around the tube after the flip is made?

19. FISHY PROBLEM

A BOWL, three-quarters filled with water, rests on a scale. If you drop a live fish into the water, the scale will show an increase in weight equal to the weight of the fish. Suppose, however, you hold the fish by its tail so that all but the extreme tip of its tail is under water. Will the scale register a greater weight than it did before you dunked the fish?

20. BICYCLE PARADOX

A ROPE is tied to the pedal of a bicycle as shown in Figure 79. If someone pulls back on this rope while another person holds the

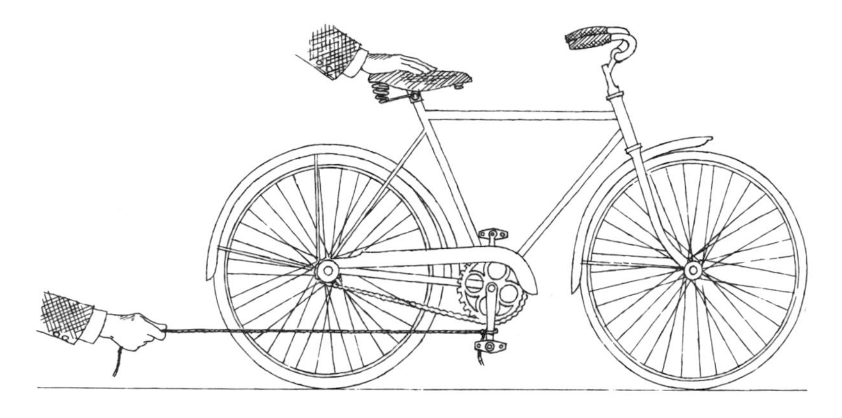

FIGURE 79
A problem in mechanics

seat lightly to keep the bicycle balanced, will the bicycle move forward, backward, or not at all?

21. INERTIAL DRIVE

If a rope is tied to the stern of a rowboat, is it possible for a man standing in the boat to propel it forward through quiet water by jerking on the free end of the rope? Could a space capsule drifting in interplanetary space be propelled by a similar method?

22. WORTH OF GOLD

Which is worth more, a pound of $10 gold pieces or half a pound of $20 gold pieces?

23. SWITCHING PARADOX

An entertaining curiosity consists of two small 110-volt bulbs (preferably one clear and one frosted) and two on-off switches

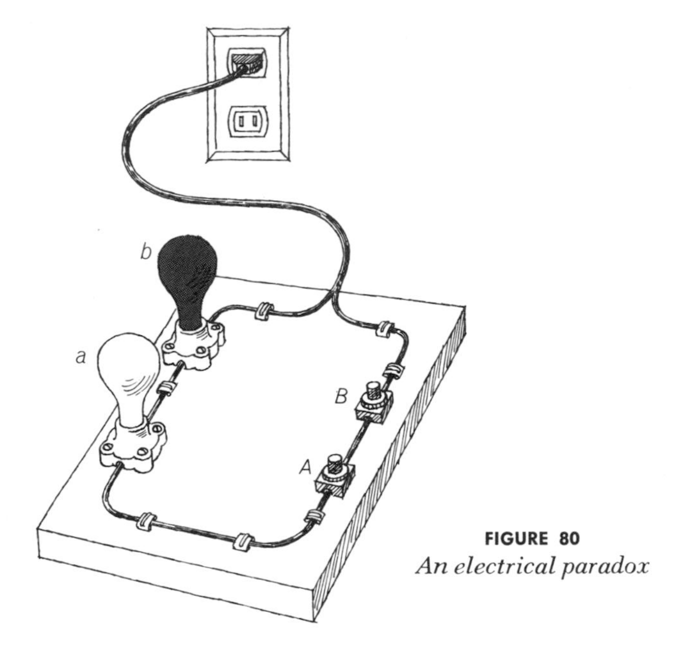

FIGURE 80
An electrical paradox

connected in a simple series circuit plugged into any wall outlet carrying the usual alternating current [*see Figure 80*]. When both switches are on, both bulbs light. If one bulb is unscrewed, the other goes out—as would be expected. When both switches are off, both bulbs are dark. But when switch *A* is on and *B* is off, only bulb *a* lights. And when switch *B* is on and *A* off, only bulb *b* lights. In short, each switch independently controls its corresponding bulb. Even more inexplicable is the fact that if the two bulbs are interchanged, switch *A* still controls bulb *a* and switch *B* still controls bulb *b*. Nothing is concealed in the wooden board on which the switches, bulb sockets, and wire are mounted. What is the secret behind the construction of this circuit?

ANSWERS

1. THE TRUCK driver is wrong. The weight of a closed compartment containing a bird is equal to the weight of the compart-

ment plus the bird's weight except when the bird is in the air and has a vertical component of motion that is accelerating. Downward acceleration reduces the weight of the system, upward acceleration increases it. If the bird is in free fall, the system's weight is lowered by the full weight of the bird. Horizontal flight, maintained by wing-flapping, alternates small up and down accelerations. Two hundred birds, flying about at random inside the panel truck, would cause minute, rapid fluctuations in weight, but the overall weight of the system would remain virtually constant.

2. When the sand is at the top of the hourglass, a high center of gravity tips the hourglass to one side. The resulting friction against the side of the cylinder is sufficient to keep it at the bottom of the cylinder. After enough sand has flowed down to make the hourglass float upright, the loss of friction enables it to rise.

If the hourglass is a trifle heavier than the water it displaces, the toy operates in reverse. That is, the hourglass normally rests at the bottom of the cylinder; when the cylinder is turned over, the hourglass stays at the top, sinking only after the transfer of sand has eliminated the friction. Shops in Paris carry the toy in both versions and in a combined version with two cylinders side by side so that as the hourglass goes up in one it sinks in the other.

The toy, said to have been invented by a Czechoslovakian glassblower, who makes them in a shop just outside Paris, is usually more puzzling to physicists than to other people. A common explanation advanced by physicists is that the force of the falling sand grains keeps the hourglass on the bottom, or at least contributes to doing so. It is not hard to show, however, that the net weight of the hourglass remains the same as if the sand were not pouring. See "Weight of an Hourglass," by Walter P. Reid, *American Journal of Physics,* Vol. 35, April, 1967, pages 351–352.

3. When an iron doughnut expands with heat, it keeps its proportions, therefore the hole also gets larger. The principle is at work when an optician removes a lens from a pair of glasses by heating the frame or a housewife heats the lid of a jar to loosen it.

4. Insert toothpick *A* between the cardboard horseshoe and toothpick *B* and move the horseshoe just enough to let the end of toothpick *B* come to rest on toothpick *A* [*see Figure 81*]. Maneuver the end of *B* under the horseshoe and then lift shoe and toothpick, balancing them as shown in the bottom drawing.

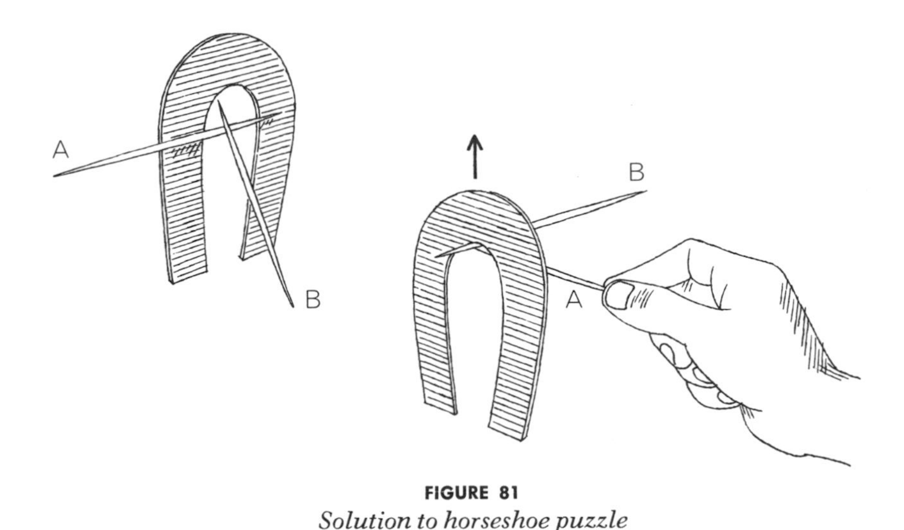

FIGURE 81
Solution to horseshoe puzzle

5. The cork floats at the center of the surface of the water in the glass only when the glass is filled a bit above its brim. The water's surface tension maintains the slightly convex surface.

Gene Lindberg and M. H. Greenblatt suggested a second method. Rotate a partially filled glass around its vertical axis. This creates a concave surface which centers the cork at the bottom. An even simpler way to create the vortex is by stirring the water with a spoon.

6. Oil floats on vinegar. To pour the oil Mr. Smith had only to tip the bottle. To pour vinegar he corked the bottle, inverted it, then loosened the cork just enough to let the desired amount of vinegar dribble out.

7. On Lewis Carroll's "boat carriage," each pair of oval wheels, on opposite sides of the same axle, turns so that at all times the long axes of the ovals are at right angles to each other. This produces the "roll." If on each side of the carriage the two wheels also had their long axes at right angles, the carriage would neither pitch nor roll. It would simply move up and down, first on two diagonally opposite wheels, then on the other two. However, by gearing the front and back wheels so that on each side of the carriage the two wheels have their long axes at 45-degree angles, the carriage can be made to pitch and roll nicely, with single wheels leaving the ground in a four-beat sequence that is repeated as the carriage moves forward.

Maya and Nicolas Slater wrote from London to say that if one abandons Lewis Carroll's proviso that elliptical wheels opposite each other must have their major axes at right angles, there is a way to make the carriage pitch and roll without any wheel's leaving the ground. The wheels must be geared so that *diagonally* opposite wheels keep their major axes at right angles. Regardless of the angle between the two front wheels, all four wheels remain on the ground at all times. If the front wheels are at a 90-degree angle, the carriage rolls without pitching; if the angle is zero, it pitches without rolling. All intermediate angles combine rolling and pitching. "Our preference is for 45 degrees," the Slaters wrote. "Our only problem is keeping our coachman."

8. Touch the end of one bar to the middle of the other. If there is magnetic attraction, the touching end must be on the magnetized bar. If not, it is on the unmagnetized bar.

9. The water level stays the same. An ice cube floats only because its water has expanded during crystallization; its weight

remains the same as the weight of the water that formed it. Since a floating body displaces its weight, the melted ice cube will provide the same amount of water as the volume of water it displaced when frozen.

10. The acrobat first ties the lower ends of the ropes together. He climbs rope *A* to the top and cuts rope *B*, leaving enough rope to tie into a loop. Hanging in this loop with one arm through it, he cuts rope *A* off at the ceiling (taking great care not to let it fall!) and then passes the end of *A* through the loop and pulls the rope until the middle of the tied-together ropes is at the loop. After letting himself down this double rope he pulls it free of the loop, thereby obtaining the entire length of *A* and almost all of *B*.

Many alternate solutions for the rope-stealing problem were received; some made use of knots that could be shaken loose from the ground, others involved cutting a rope partway through so that it would just support the thief's weight and later could be snapped by a sudden pull. Several readers doubted that the thief would get *any* rope because the bells would start ringing.

11. The top of the shadow of a man walking past a street light moves faster than the man, but it maintains a constant speed regardless of its length.

12. A quantity of water flows over the first winding of the hose to fall to the bottom and form an air trap. The trapped air prevents any more water from entering the first loop of the hose.

If the funnel end of the empty hose is high enough, water poured into it will be forced over more than one winding to form a series of "heads" in each coil. The maximum height of each head is about equal to the coil's diameter. The diameter, times the number of coils, gives the approximate height the water column at the funnel end must be to force water out at the other end. (This was pointed out by John C. Bryner, Jan Lundberg, and J. M. Osborne.) W. N. Goodwin, Jr., noted that for hoses with an outside diameter of 5/8 inch or less the funnel end can be as low as twice the height of the coils and water will flow all the way through a series of many coils. The reason for this is not yet clear.

13. To remove the egg, tip back your head, hold the bottle upside down with the opening at your lips and blow vigorously into it. When you take the bottle from your mouth, the compressed air will pop the egg out through the neck of the bottle.

A common misconception about this trick is that the egg was initially drawn into the bottle by a vacuum created by the loss of oxygen. Oxygen is indeed used up, but the loss is compensated by the production of carbon dioxide and water vapor. The vacuum is created solely by the quick cooling and contracting of the air after the flames go out.

14. The nuts and bolts inside the toy ship displace an amount of water equal to their *weight*. When they sink to the bottom of the tub, they displace an amount equal to their *volume*. Since each piece weighs considerably more than the same volume of water, the water level in the tub is lowered after the cargo is dumped.

15. As the closed car accelerates forward, inertial forces send the air in the car backward. This compresses the air behind the balloon, pushing it forward. As the car rounds a curve, the balloon, for similar reasons, moves *into* the curve.

16. Zero gravity prevails at all points inside the hollow asteroid. For an explanation of how this follows from gravity's law of inverse squares see Hermann Bondi's Anchor paperback *The Universe at Large*, page 102.

H. G. Wells, in his *First Men in the Moon*, failed to realize this. At two points in the novel he has his travelers floating near the center of their spherical spaceship because of the gravitational force exerted by the ship itself; this aside from the fact that the gravity field produced by the ship would be too weak to influence the travelers anyway.

17. A bird cannot fly at all on the moon because there is no lunar air to support it. George Milwee, Jr., wrote to ask if I got the idea for this problem from Immanuel Kant! In the Introduction to *The Critique of Pure Reason,* section 3, Kant chides Plato for thinking that he can make better philosophical progress if he abandons the physical world and flies in the empty space of pure reason. Kant bolsters his point with the following delightful metaphor: "The light dove, cleaving the air in her free flight, and feeling its resistance, might imagine that its flight would be still easier in empty space."

18. Consider any particle k in the liquid, before the tube is flipped. The earth carries it counterclockwise (looking down at the North Pole) at a speed of one revolution in 24 hours. But every point in the tube, the same distance from the earth's center, is revolving counterclockwise at the same speed as the particle, therefore the circulatory speed of k, relative to the tube, is zero. After the flip, k continues counterclockwise at the same speed, but the tube's reversal gives k a clockwise motion, relative to the tube, that carries it once around the tube (ignoring friction) in exactly half the time it takes the earth to complete one rotation; namely, 12 hours.

This curious halving of 24 hours can be better understood by considering a specific particle, say one at the top of the tube. Let x be the earth's diameter and y the tube's diameter. Both the particle and the tube's top travel east at a speed of $\pi(x + 2y)$ in 24 hours. After the flip, the particle continues at the same speed, but now it is at the bottom where the tube moves east at the slower rate of πx in 24 hours. Consequently, the particle, relative to the tube, moves east at a speed of $\pi(x + 2y) - \pi x = 2\pi y$

in 24 hours. Since πy is the tube's circumference, we see that the particle is going around the tube at a speed of two circuits in 24 hours, or one in 12. Similar calculations apply to all other particles in the tube.

Most people, in calculating the ideal speed of the water, miss by a factor of 2 and assume it would circulate once around the tube in 24 hours. Compton himself made this error, both in his first paper (*Science*, Vol. 37, May 23, 1913, pages 803–806), which describes work with a tube of one-meter radius, and a second paper, reporting results with more elaborate apparatus (*Physical Review*, Vol. 5, February, 1915, page 109, and *Popular Astronomy*, Vol. 23, April, 1915, page 199).

When the tube is not at the equator, the maximum effect is achieved by having the plane of the tube tilted from the vertical before the 180-degree flip is made. The nearer to the pole, the greater the tilt needed, which makes it possible to use the tube for determining its latitude. At the North or South Pole, the greatest effect is obtained with the tube horizontal before it is flipped. In all latitudes, when the plane of the tube is properly tilted and oriented for maximum effect, the flip induces a flow of one revolution in 12 hours, assuming of course a total absence of viscosity.

The full story is somewhat complicated, and there is an interesting correspondence between what happens in the Compton tube and the induction of electrical currents that circulate within a conducting ring when it is flipped 180-degrees in a magnetic field. I am indebted to Dave Fultz, of the Hydrodynamics Laboratory of the University of Chicago's department of geophysical sciences, for supplying the information on which this brief account is based. Fultz prefers an explanation in terms of angular velocities and torques, but since this gets into difficult technical terms and concepts, I have let the explanation stand as it is, in the simpler but cruder terms of linear velocities.

When I was an undergraduate at the University of Chicago, I attended a lecture by Compton in which he spoke of his experiment. Afterward, I approached him and asked if it would be possible to build up a stronger circulation in the tube by alternately flipping it around horizontal and vertical axes. Looking puzzled, Compton took a half dollar out of his pocket and began turning it between thumb and finger, muttering to himself. He finally shook his head and said he believed it would not work, but that he would think about it.

19. The bowl increases in weight by the amount of liquid displaced by the dunked fish.

20. Pulling the lower pedal of the bicycle backward causes the pedal to rotate in a way that normally would drive the bicycle forward, but since the coaster brake is not being applied the bicycle is free to move back with the pull. The large size of the wheels and the small gear ratio between the pedal and the wheel sprockets is such that the bicycle moves backward. The pedal moves back also with respect to the ground, although it moves forward with respect to the bicycle. When it rises high enough, the brake sets and the bicycle stops. Readers who do not believe all this will simply have to get a bicycle and try it. The apparent paradox is explained in many old books. For a recent analysis, see D. E. Daykin, "The Bicycle Problem," *Mathematics Magazine*, Vol. 45, January, 1972, p. 1.

21. A rowboat *can* be moved forward by jerking on a rope attached to its stern. In still water a speed of several miles per hour can be obtained. As the man's body moves toward the bow, friction between boat and water prevents any significant movement of the boat backward, but the inertial force of the jerk is strong enough to overcome the resistance of the water and transmit a forward impulse to the boat. The same principles apply when a boy sits inside a carton and scoots himself across a waxed floor by rapid forward movements inside the box. No such "inertial space drive" is possible inside a spaceship because the near vacuum surrounding the ship offers no resistance.

22. A pound of $10 gold pieces contains twice as much gold as half a pound of $20 gold pieces, therefore it is worth twice as much.

23. Concealed inside the base of each bulb and the base of each on-off switch is a tiny silicon rectifier that allows the current to flow through it in only one direction. The circuit is shown in Figure 82, the arrows indicating the direction of current flow permitted by each rectifier. If the current is moving so that a rectifier in the base of a bulb will carry it, the rectifier steals the current and the bulb remains dark. It is easy to see that each switch turns on and off only the bulb whose rectifier points in the same circuit direction as the rectifier in the switch.

Aside from complaints about suddenly tossing in diodes (rectifiers) from left field, readers also complained (understandably) of the difficulty of inserting a tiny diode in the base of a bulb. R. Allen Pelton found it easier to make a working model as follows: "I connected the diode under the base of the porcelain light socket, then cut out the wood base to hold it. I am unable to switch the bulbs, but then again I can insert *any* bulb. My little wood panel has stumped every person yet I have shown it to."

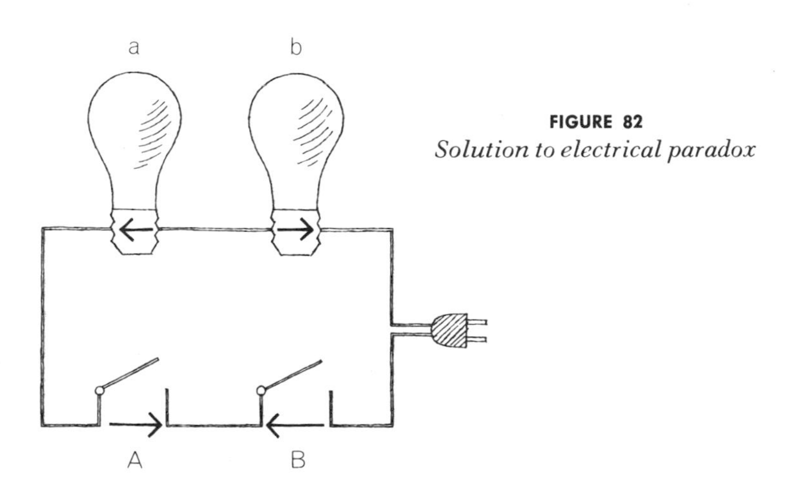

FIGURE 82
Solution to electrical paradox

CHAPTER 15

Pascal's Triangle

There are so many relations present [in Pascal's triangle] that when someone finds a new identity, there aren't many people who get excited about it anymore, except the discoverer!

—DONALD E. KNUTH, in *Fundamental Algorithms*

HARRY LORAYNE, a professional magician and memory expert who lives in New York City, likes to puzzle friends with an unusual mathematical card trick. A spectator is given a deck from which the face cards and tens have been removed. He is asked to place any five cards face up in a row. Lorayne immediately finds a card in the deck that he puts face down at a spot above the row, as shown in Figure 83. The spectator now builds a pyramid of cards as follows:

Each pair of cards in the row is added by the process of "casting out nines." If the sum is above 9, 9 is subtracted. This can be done rapidly by adding the two digits in the sum. For example, the first two cards in the bottom row of the illustration add to 16. Instead of subtracting 9 from 16, the same result is obtained by adding 1 and 6. The sum is 7; therefore the spectator puts a seven above the first pair of cards. The second and third cards add to 8, so an eight goes above them. This is continued until a new row of four cards is obtained, and the procedure is repeated until the pyramid reaches the face-down apex card. When this card is turned over, it proves to be the correct value for the final sum.

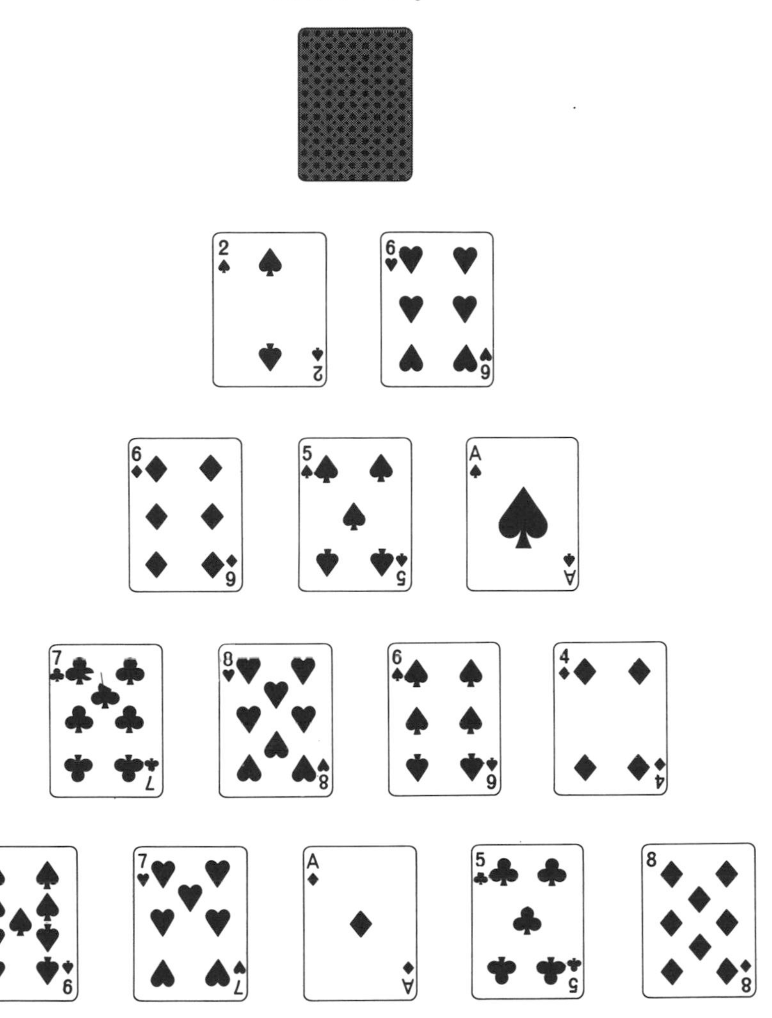

FIGURE 83
The apex card trick

The trick can be done with any number of cards in the initial row, although if there are too many there may not be enough cards to supply all the needed values for the pyramid. The computations can, of course, always be done on paper. A good version of the trick is to ask someone to jot down a row of 10 random digits. You can calculate the pyramid's apex digit quickly in your head if you know the secret, and it will always turn out to be correct. How is the apex digit determined? One's first thought is that perhaps it is the "digital root" of the first row—the sum of the digits reduced to a single digit by casting out nines—but this is not the case.

The truth is that Lorayne's trick operates with simple formulas derived from one of the most famous number patterns in the history of mathematics. The pattern is known as Pascal's triangle because Blaise Pascal, the 17th-century French mathematician and philosopher, was the first to write a treatise about it: *Traité du triangle arithmétique* (*Treatise on the Arithmetic Triangle*). The pattern was well known, however, long before 1653, when Pascal first wrote his treatise. It had appeared on the title page of an early 16th-century arithmetic by Petrus Apianus, an astronomer at the university in Ingolstadt. An illustration in a 1303 book by a Chinese mathematician also depicts the triangular pattern, and recent scholarship has traced it back still earlier. Omar Khayyám, who was a mathematician as well as a poet and philosopher, knew of it about 1100, having in turn probably got it from still earlier Chinese or Indian sources.

The pattern is so simple that a 10-year-old can write it down, yet it contains such inexhaustible riches, and links with so many seemingly unrelated aspects of mathematics, that it is surely one of the most elegant of all number arrays. The triangle begins with 1 at the apex [*see Figure 84*]. All other numbers are the sums of the two numbers directly above them. (Think of each 1, along the two borders, as the sum of the 1 above it on one side and 0, or no number, on the other.) The array is infinite and bilaterally symmetric. In the illustration the rows and diagonals are numbered in the customary way, beginning with 0 instead of 1, to simplify explaining some of the triangle's basic properties.

Diagonal rows, parallel to the triangle's sides, give the triangular numbers and their analogues in spaces of all dimen-

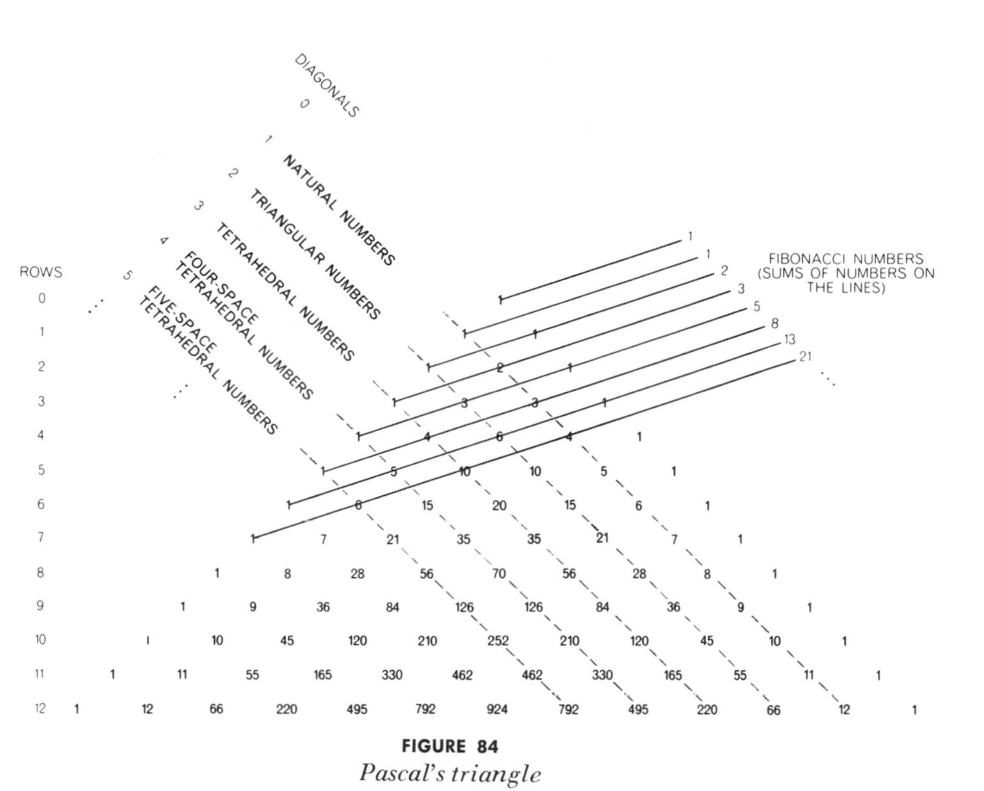

FIGURE 84
Pascal's triangle

sions. A triangular number is the cardinal number of a set of points that will form a triangular array. This sequence of triangular numbers (1, 3, 6, 10, 15 . . .) is found in the triangle's second diagonal. (Note that every adjacent pair of numbers adds to a square number.) The first diagonal, consisting of natural numbers, gives the analogues of triangular numbers in one-dimensional space. The zero diagonal gives the analogue in zero-space, where the point itself is obviously the only possible pattern. The third diagonal contains tetrahedral numbers: cardinal numbers of sets of points that form tetrahedral arrays in three-space. The fourth diagonal gives the number of points that form hypertetrahedral arrays in four-space, and so on for the infinity of other diagonals. The nth diagonal gives the n-space analogues of triangular numbers.

We can see at a glance that 10 cannonballs will pack into a tetrahedral pyramid or into a flat triangle, and that the 56 hypercannonballs in a five-space tetrahedron can be rearranged on a hyperplane to form a tetrahedron (but if we try to pack them on a plane in triangular formation, there will be one left over).

To find the sum of all the numbers in any diagonal, down to any place in the series, simply look at the number directly below and left of the last number in the series to be summed. For example, what is the sum of the natural numbers from 1 through 9? Move down the first diagonal to 9, then down and left to 45, the answer. What is the sum of the first eight triangular numbers? Find the eighth number in the second diagonal, move down and left to 120, the answer. If we put together all the balls needed to make the first eight triangles, they will make exactly one tetrahedral pyramid of 120 balls.

The sums of the more gently sloping diagonals, indicated by solid lines, form the familiar sequence of Fibonacci numbers, 1, 1, 2, 3, 5, 8, 13 . . . , in which each number is the sum of the two numbers preceding it. (Can you see why?) The Fibonacci sequence often turns up in combinatorial problems. To cite one instance, consider a row of n chairs. In how many different ways can you seat men and women in the chairs provided that no two women are allowed to sit next to each other? When n is 1, 2, 3, 4 . . . , the answers are 2, 3, 5, 8 . . . and so on in the Fibonacci order. Pascal apparently did not know that the Fibonacci series was embedded in the triangle; it seems not to have been noticed until late in the 19th century.

And not until recently was it noticed that by removing diagonals from the left side of the triangle one obtains partial sums for the Fibonacci series. The discovery was made by Verner E. Hoggatt, Jr., a mathematician at San Jose State College who edits *The Fibonacci Quarterly*, a fascinating journal that has published many articles about Pascal's triangle. If the zero diagonal on the left side is sliced off, the Fibonacci diagonals have sums that are the partial sums of the Fibonacci series ($1 = 1$; $1 + 1 = 2$; $1 + 1 + 2 = 4$; $1 + 1 + 2 + 3 = 7$; and so on). If diagonals 0 and 1 are eliminated from the left side, the Fibonacci diagonals give the partial sums of the partial sums ($1 = 1$; $1 + 2 = 3$; $1 + 2 + 4 = 7$; and so forth). In general, if k diagonals are trimmed, the Fibonacci diagonals give the k-fold partial sums of the Fibonacci series.

Each horizontal row of Pascal's triangle gives the coefficients in the expansion of the binomial $(x + y)^n$. For example, $(x + y)^3 = x^3 + 3x^2y + 3xy^2 + y^3$. The coefficients of this expansion are 1, 3, 3, 1 (a coefficient of 1 is customarily omitted from a term), which is the third row of the triangle. To find the coefficients of $(x + y)^n$, in proper order, merely look at the triangle's nth row. This basic property of the triangle ties it in with elementary combinatorial and probability theory in ways that make the triangle a useful calculating device. Suppose an Arab chief offers to give you any three of his seven wives. How many different selections can you make? You have only to find the intersection of diagonal 3 and row 7 to get the answer: 35. If (in your eager confusion) you commit the blunder of looking for the intersection of diagonal 7 and row 3, you will find that they do not intersect, so that the method can never go wrong. In general the number of ways to select a set of n elements from a set of r distinct elements is given by the intersection of diagonal n and row r.

The connection between this and probability is easily seen by considering the eight equally possible outcomes of getting heads or tails when flipping three pennies: HHH, HHT, HTH, HTT, THH, THT, TTH, TTT. There is one way to get three heads, three ways to get two heads, three ways to get one head and one way to get no heads. These numbers (1, 3, 3, 1) are, of course, the triangle's third row. Suppose you want to know the probability of exactly five heads showing if you toss 10 pennies in the air. First determine how many different ways five pennies can be selected from 10. The intersection of diagonal 5 and row 10 provides the answer: 252. Now you must add the numbers in the 10th row to obtain the number of equally possible cases. You can short-cut this addition by remembering that the sum of the nth row of Pascal's triangle is always 2^n. (The sum of each row is obviously twice the sum of the preceding row, since every number is carried down twice to enter into the numbers of the row below; therefore the sums of the rows form the doubling series 1, 2, 4, 8. . . .) The 10th power of 2 is 1,024. The probability of getting five heads is 252/1,024, or 63/256. (There is a mechanical device for demonstrating probability, often exhibited at science fairs and museums, in which hundreds of small balls roll down an incline through a hexagonal array of obstacles to enter slots and form an approximation of the bell-shaped normal-distribution curve. For a picture of such a device, and a discussion of how Pascal's triangle underlies it, see "Probability," by Mark Kac, in *Scientific American*, September, 1964.)

If we represent each number of the triangle by a small dot, then blacken every dot whose number is not exactly divisible by a certain positive integer, the result is always a striking pattern of triangles. Patterns obtained in this way conceal many surprises. Consider the binary pattern that results when the divisor is 2 [*see Figure 85*]. Running down the center are gray triangles of increasing size, each made up entirely of even-numbered dots. At the top is a "triangle" of one dot, then the series continues with triangles of 6, 28, 120, 496 . . . dots. Three of those numbers—6, 28, and 496—are known as perfect numbers because each is the sum of all its divisors, excluding itself (for example, $6 = 1 + 2 + 3$). It is not known if there is an infinity of perfect numbers, or if there is one that is odd. Euclid managed to prove, however, that every number of the form $2^{n-1}(2^n - 1)$, where $(2^n - 1)$ is a prime, is an even perfect number. Leonhard Euler much later showed that all even perfect numbers conform to Euclid's formula. The formula is equivalent to

$$\frac{P(P+1)}{2},$$

where P is a Mersenne prime (a prime that has the form $2^p - 1$, where p is a prime). The above expression happens also to be

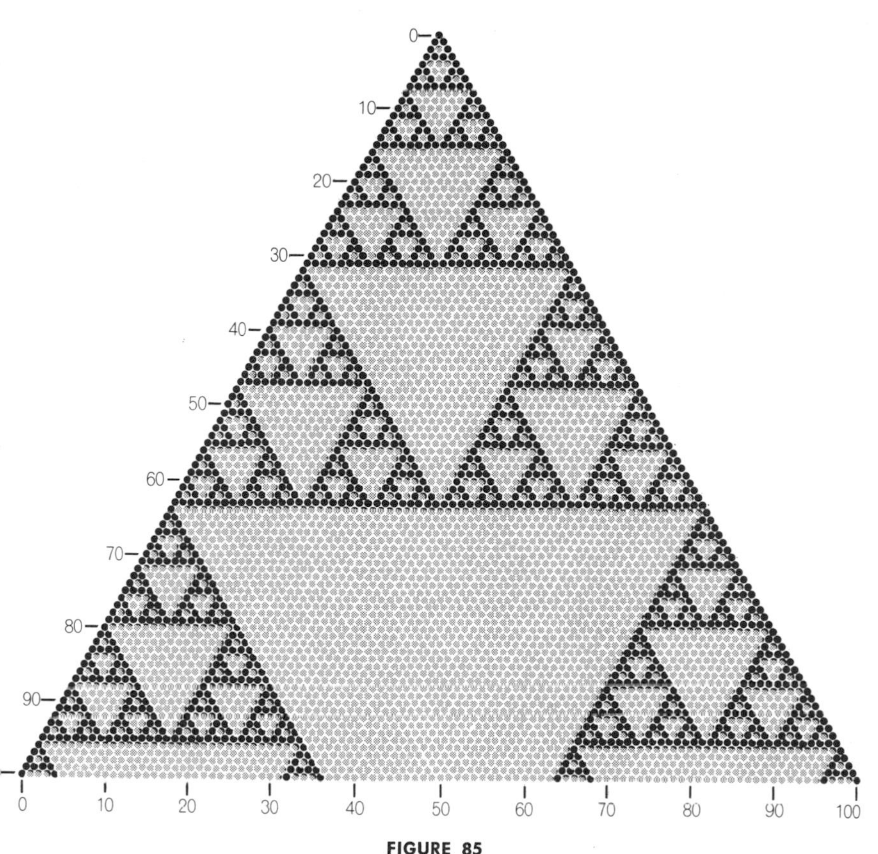

FIGURE 85

Pascal's triangle with numbers represented by dots—the odd numbers by black dots

the formula for a triangular number. In other words, if the "side" of a triangular number is a Mersenne prime, the triangular number is also perfect. Going back to the even-odd coloring of Pascal's triangle, it can be shown that the formula for the number of dots in the nth central triangle, moving down from the apex, is $2^{n-1}(2^n - 1)$, the formula for perfect numbers. All even perfect numbers appear in the pattern, therefore, as the number of dots in the nth central triangle whenever $2^n - 1$ is a prime. Because $2^4 - 1 = 15$, which is not a prime, the fourth gray triangle is not perfect. The fifth triangle of 496 dots *is* perfect because $2^5 - 1 = 31$, a prime. (The sixth gray triangle is not perfect, but the seventh, with 8,128 dots, is.)

One final curiosity. If rows 0 through 4 are read as single numbers (1, 11, 121, 1,331, and 14,641), they are the first five powers of 11, starting with $11^0 = 1$. The fifth row *should* be $11^5 = 161{,}051$, but it is not. Observe, however, that this is the first row with two-digit numbers. If we interpret each number as indicating a multiple of the place value of that spot in decimal notation, the fifth row can be interpreted (reading right to left) as $(1 \times 1) + (5 \times 10) + (10 \times 100) + (10 \times 1{,}000) + (5 \times 10{,}000) + (1 \times 100{,}000)$, which gives the correct value of 11^5. Interpreted this way, each nth row is 11^n. [For three articles on Pascal's triangle and powers of 11, see *Mathematics Teacher,* Vol. 57 (1964), page 392; Vol. 58 (1965), page 425; and Vol. 59 (1966), page 461.]

Almost anyone can study the triangle and discover more properties, but it is unlikely they will be new, for what is said here only scratches the surface of a vast literature. Pascal himself, in his treatise on the triangle, said that he was leaving out more than he was putting in. "It is a strange thing," he exclaimed, "how fertile it is in properties!" There are also endless variants on the triangle, and many ways to generalize it, such as building it in tetrahedral form to give the coefficients of trinomial expansions.

If the reader can solve the following five elementary problems, he will find his understanding of the triangle's structure pleasantly enriched:

1. What formula gives the sum of all numbers above row n? (Rows are numbered as in Figure 84, starting with zero for the apex number.)

2. How many odd numbers are there in row 255?

3. How many numbers in row 67 are evenly divisible by 67?

4. If a checker is placed on one of the four black squares in the first row of an otherwise empty checkerboard, it can move (by standard checker moves) to any of the four black squares on the last (eighth) row by a variety of different paths. One pair of starting and ending squares is joined by a maximum number of different routes. Identify the two squares and give

the number of different ways the checker can move from one to the other.

5. Given an initial row of n cards, in the pyramid trick described at the beginning, how can one obtain from Pascal's triangle simple formulas for calculating the value of the apex card?

ADDENDUM

ANSWER 5, in the section to follow, tells how Pascal's triangle is used for solving the pyramid trick. To understand why the formula works, consider the triangle shown in Figure 83, and assume that adjacent cards are summed without casting out nines. The pyramid will be:

```
                71
            38      33
        24      14      19
    16       8       6      13
 9       7       1       5       8
```

The key row of Pascal's triangle is 1,4,6,4,1. The 1's at its ends tell us that, as we move upward making additions, the value of each end card on the bottom row enters only once into the final summation. That is because there is only one path from each of these cards to the top. The 4's that are second from the ends of the key tell us that the value of each card second from an end enters four times into the final sum because there are four forking paths from each of these cards to the top. The central 6 of the key tells us that there are six forking paths from the center card to the top, therefore the value of this card enters six times into the final sum. Accordingly: $(1 \times 9) + (4 \times 7) + (6 \times 1) + (4 \times 5) + (1 \times 8) = 71$, the apex number. Since this procedure gives the sum at the apex, it must also provide the apex digital root if nines are cast out.

Magicians know the card trick as "Apex." It was originated by a German magician, Franz Braun, who published it about 1960 in his regular column on mathematical tricks in *Magie*, a

German magic periodical. See Ronald Wohl's note on this in *The Pallbearers Review* (an American magic journal), June, 1967, page 105.

When the trick is done with cards it is good to have a second deck handy in case the pyramid requires more than four cards of the same value. This can happen even with small pyramids. For example, a bottom row of 4,5,4,5 would require six nines to complete the structure.

C. J. H. Wevers, a reader in Holland, posed an interesting problem. If we remove the face cards and the tens from a normal deck, just 36 cards are left, and 36 is a triangular number. Is it possible, Wevers asked, to form a row of eight of these cards such that, after the triangle is completed according to the rules of Apex, just those 36 cards will be used? "It is clear," Wevers wrote, "that it is not an easy problem to solve, if it can be solved. I am convinced that it should be rather simple to make a computer program for it. . . ."

I found that the problem has one elegant solution. The reader may enjoy searching for it. Not counting reversals as different, is there more than one solution?

ANSWERS

1. THE SUM of all numbers above row n is $2^n - 1$.

2. All numbers in row n are odd if, and only if, n is a power of 2 diminished by 1. Because $255 = 2^8 - 1$, all its numbers are odd.

3. All numbers in row 67, except the two 1's at the ends, are exactly divisible by 67. All numbers in a row n are exactly divisible by n if, and only if, n is a prime. A proof will be found in Stanley Ogilvy's *Through the Mathescope*, page 137.

4. The checker problem is quickly solved by numbering the squares as shown in Figure 86. For each starting position the

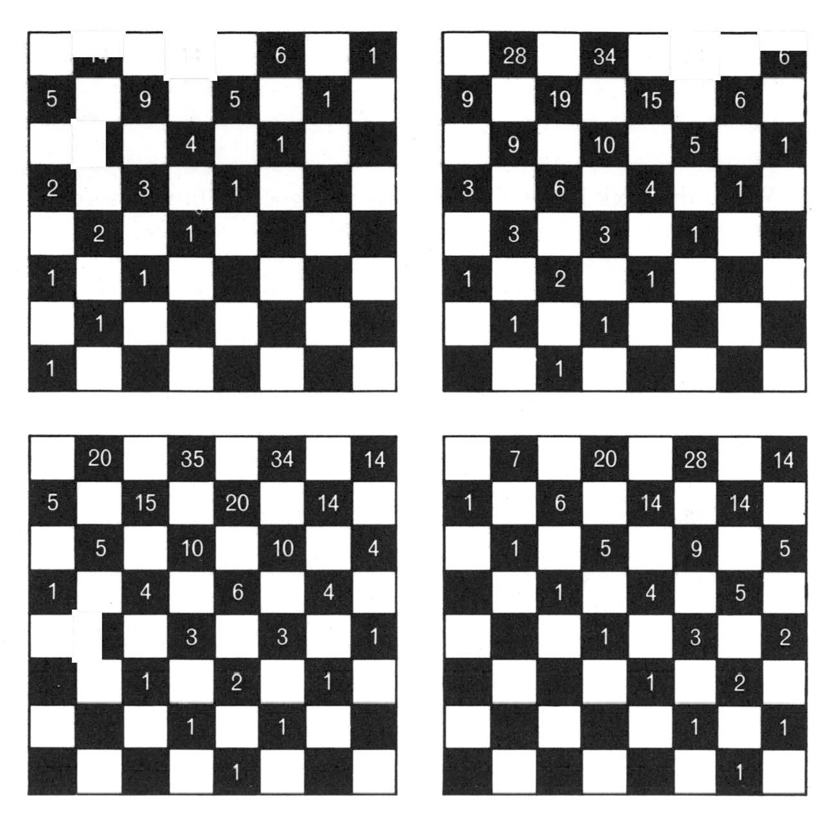

FIGURE 86
Solution to checker problem

numbers form inverted Pascal triangles modified by the restricting sides of the board. Each number indicates the number of different ways a checker can reach that cell from the starting position. The maximum number of possible paths is to the square marked 35 when the checker starts from the third black square on the bottom row.

5. The value of the apex card in Harry Lorayne's trick is determined as follows. Let n be the number of cards in the initial row. The row of Pascal's triangle that contains n numbers provides the formula for calculating the apex. This can be explained by some examples.

Assume that there are six cards in the bottom row with the values 8, 2, 9, 4, 6, 7. The corresponding row of Pascal's triangle is 1, 5, 10, 10, 5, 1. Reduce the 10's to their digital root (by adding their digits), making the row 1, 5, 1, 1, 5, 1. These numbers are taken as being multiples of the six cards. The cards that are second from each end are multiplied by 5, summed, then added to the values of the four remaining cards. The final sum, reduced to its digital root, is the apex. This is easily done in the head because you can reduce to digital roots as you go along. When the second-from-end cards are multiplied by 5 to obtain the numbers 10 and 30, those numbers are immediately reduced to their digital roots 1 and 3, which have a sum of 4. To 4 you now add the values of the remaining four cards, reducing each sum to its digital root as you proceed. The final result, 5, is the apex number.

For the pyramid shown in Figure 83, with five cards in the bottom row, the fifth row of Pascal's triangle provides the key: 1, 4, 6, 4, 1. The apex card is the digital root of the sum of the bottom cards after the center card has been multiplied by 6 and each of its neighbors by 4. This calls for more head work on your part than a pyramid with a six-card base, but requires less work on the spectator's part. Incidentally, eliminating face cards and tens from the deck is also done solely to simplify the spectator's calculations. The trick works just as well if the entire deck is used, with values of 11, 12, and 13 assigned to jacks, queens, and kings.

Finding the apex is easiest for a row of 10 numbers. In this case the corresponding row of Pascal's triangle, reduced to digital roots, is 1, 9, 9, 3, 9, 9, 3, 9, 9, 1. The number 9 is the same as 0 (modulo 9), so that we can write the formula: 1, 0, 0, 3, 0, 0, 3, 0, 0, 1. To obtain the apex, therefore, we have only to multiply the fourth numbers from each end by 3, add the two end cards, and reduce to the digital root. The other six numbers can be ignored completely! A good presentation (suggested by L. Vosburgh Lyons) is to let the spectator himself predict the apex number by naming any digit he pleases. He then writes a row of nine random digits, allowing you to add a 10th digit at whichever end of the row he designates. Add the three key numbers of the formula in the usual manner, then supply whatever fourth number is needed to make the apex correspond with his prediction.

The trick need not be limited to "casting out nines" addition. Any integer may be cast out. Pascal's triangle, with its numbers reduced by the same kind of casting out, gives the required formulas. For example, suppose the trick starts with eight digits and the pyramid is formed by casting out sevens. The eight-number row of Pascal's triangle, reduced by casting out sevens, is 1, 0, 0, 0, 0, 0, 0, 1. To determine the apex merely add the end numbers and, if necessary, reduce to a digit by casting out seven. I leave it to the reader to determine why the triangle produces the desired formulas in all such cases.

CHAPTER 16

Jam, Hot, and Other Games

IN THIS chapter we consider a variety of two-person games, some old and some new, for which mathematical strategies are known. First, here is a trio of simple games that are related to each other in an amusing and surprising way.

1. Nine playing cards, with values from ace to nine, are face up on the table. Players take turns picking a card. The first to obtain three cards that add to 15 is the winner.

2. On the road map in Figure 87, players take turns eliminating one of the nine numbered highways. This is done by coloring the complete length of the road, even though it may go through one or two towns (*circles*). Pencils of two different colors are used to distinguish the moves of the two players. The first to color three highways that enter the same town is the winner. (The Dutch psychologist John A. Michon, who invented this game, calls it "Jam" because those are his initials and because the object of the game is to jam crossings by blocking highways.)

3. Each of the following words is printed on a card: HOT, HEAR, TIED, FORM, WASP, BRIM, TANK, SHIP, WOES. The nine cards are placed face up on the table. Players take turns removing a card. The first to hold three cards that bear the same letter is the winner. (The Canadian mathematician Leo Moser, who devised this game, called it "Hot.")

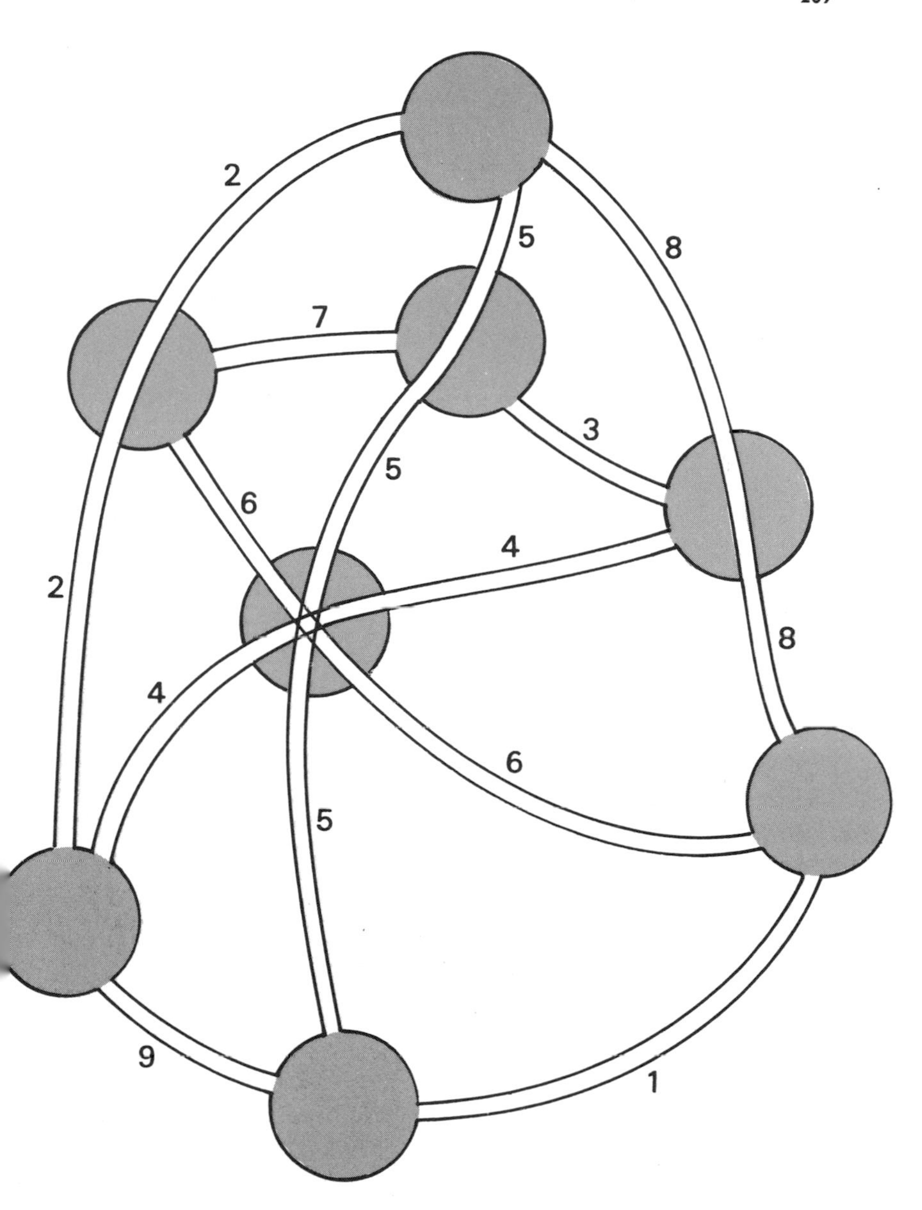

FIGURE 87
Map for the game of Jam

For each game the question is: If both players make their best moves, is the game a win for the first player, a win for the second player, or a draw? Perhaps the reader has already experienced what the Gestalt psychologists call "closure" and recognized that all three games are isomorphic with ticktacktoe!

It is easy to see that this is the case. For the first game we make a list of all the triplets of distinct digits from 1 to 9 that have a sum of 15. There are exactly eight such triplets. They can be interlocked on a ticktacktoe board as shown in Figure 88 to form the familiar order-3 magic square on which every row, column, and main diagonal is one of the triplets. Each numbered card drawn by a player corresponds to a ticktacktoe play on the cell of the magic square that bears that digit. Each set of triplets that wins in the card game corresponds to a winning ticktacktoe row on the magic square. Anyone who can play a perfect game of ticktacktoe and who also memorizes the magic square can immediately play a perfect game in this card version.

The map in Figure 87 is topologically equivalent to the symmetrical graph shown at the left of Figure 89. This is in turn the "dual" of the graph obtained by connecting the centers of the nine cells of a ticktacktoe board as shown at the right in the illustration. Each numbered cell of the magic square corresponds to a numbered highway on the map and each town on the map corresponds to a row, column, or main diagonal on the magic square. As before, there is an equivalence relation between plays on the map and plays in ticktacktoe.

The isomorphism of Moser's word game and ticktacktoe becomes obvious when the nine words are written inside the cells of a ticktacktoe matrix as shown in Figure 90. Each set of three-in-a-row words has a common letter, and there are no such sets other than the eight displayed in this way. Again, memorizing the square of words instantly enables a perfect-game ticktacktoe player to play a perfect game of Hot. Since ticktacktoe played rationally is always a draw, the same is true of the three equivalent games, although the first player naturally has a strong advantage over a second player who is not aware that he is playing disguised ticktacktoe or who may not play a perfect game of ticktacktoe.

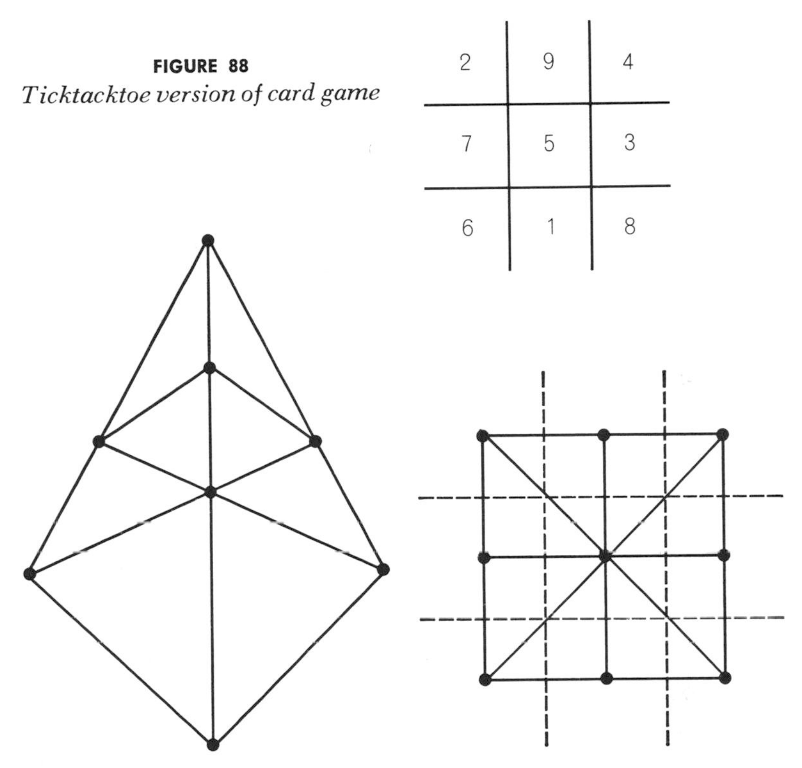

FIGURE 88
Ticktacktoe version of card game

FIGURE 89
Graph of Jam map (left) and its ticktacktoe "dual" (right)

HOT	FORM	WOES
TANK	HEAR	WASP
TIED	BRIM	SHIP

FIGURE 90
Key to the game of Hot

One who grasps the essential identity of the three games will have obtained a valuable insight; mathematics abounds with "games" that seem to have little in common and yet are merely two different sets of symbols and rules for playing the *same* game. Geometry and algebra, for example, are two ways of playing exactly the same game, as Descartes's great discovery of analytic geometry shows.

There are many games of the "take-away" type in which players alternately take away an element or subset from a set, the winner being the person who acquires the last element. The best-known game of this kind is nim, played with a set of counters arranged in an arbitrary number of rows, with an arbitrary number of counters in each row. On his turn a player may take as many counters as he wishes, provided that they all come from the same row. The person who takes the last counter wins. A perfect strategy is easily formulated in the binary system, as explained in *The Scientific American Book of Mathematical Puzzles & Diversions.*

A starting pattern for nim, as it was played throughout the French film *Last Year at Marienbad*, is shown in Figure 91. Sixteen cards are arranged in four rows of one, three, five, and seven cards. (The triangular pattern symbolizes the triangular love game played in the picture.) To determine whether the first or the second player can win we write the numbers of cards in each row in the binary system, then add the columns:

$$\begin{array}{lr} 1 & 1 \\ 3 & 11 \\ 5 & 101 \\ 7 & \underline{111} \\ & 224 \end{array}$$

If the sum of every column is an even number (or zero if the addition is made modulo 2), as in this case, the pattern is called "safe." This means that the first player is certain to lose against an expert, for regardless of how he plays he will leave an "un-

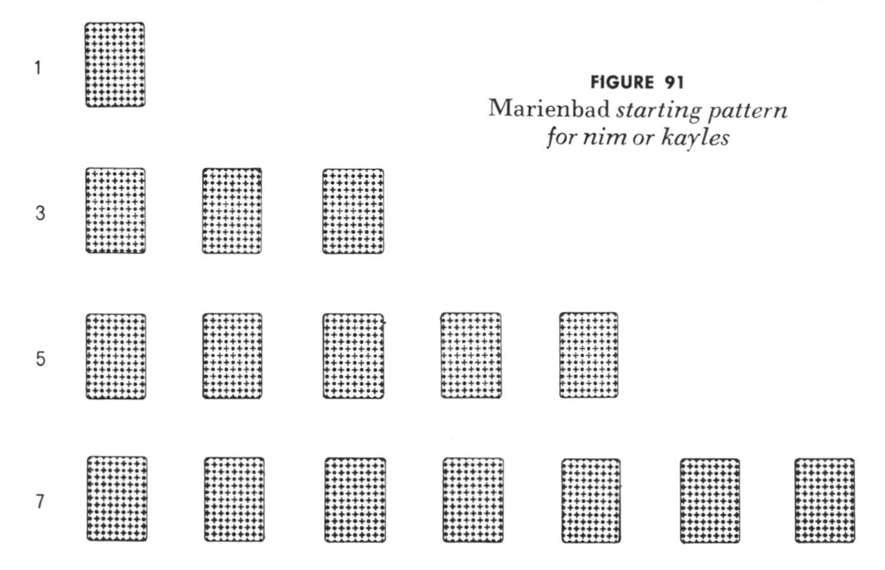

FIGURE 91
Marienbad *starting pattern for nim or kayles*

safe" pattern (one with at least one column that has an odd sum), and the second player can convert this to another safe position on his next move. By always playing to leave a safe pattern he is sure to get the last counter. (In the movie, the game is played in reverse form: the last to take *loses*. This calls for only a trivial change of strategy at the end of a game. The winner departs from normal strategy when it is possible to leave an odd number of single-counter rows.)

Michel Hénon, a mathematician at the Centre National de la Recherche Scientifique in Paris, recently thought of a delightful nim variant, played with scissors and pieces of string. It is best approached, however, by first explaining an older variant of nim called kayles, to which the string game is closely linked.

Kayles was invented by the English puzzle expert Henry Ernest Dudeney, who introduced it in Problem 73 of his first book, *The Canterbury Puzzles* (1907). It is now called kayles because Dudeney presented it as a problem that might have arisen in playing a popular 14th-century game of that name in which a ball was rolled at wooden pins standing side by side. The ball's size was such that it could knock over either a single pin or two touching pins. Players alternately roll a ball, and the person who knocks over the last pin (or pair of pins) wins.

Mathematical kayles is best played on the table with coins, cards, or other objects simply by arranging them in an arbitrary number of rows, exactly as in nim, with an arbitrary number of objects in each row. Now, however, we must think of each row as a linked chain. One may remove one link or two adjacent links. If the object or pair of objects is taken from inside a chain, it breaks the chain into two separate chains. For instance, if the first player takes the center card from the bottom row of the Marienbad pattern, it breaks the seven cards into two separate chains of three links each. In this way the number of chains is likely to increase as the game continues. The person who plays last wins.

Kayles also lends itself to binary analysis, but not as directly as nim. For every chain we associate a binary number, but that number (except for the three smallest cases) is not the same as the decimal number of the cards in the chain. The chart shown in Figure 92, supplied by Hénon, gives the required binary number, here called the k number, for integers 1 through 70. After 70 a curious periodicity of 12 numbers sets in. If the number is above 70, divide it by 12, note the remainder, then use the chart at the bottom right of Figure 92. To decide if a kayles pattern is safe or unsafe, one uses k numbers like the nim binary numbers.

Consider the Marienbad starting position, which is safe in nim and therefore a win for the second player. Is it also safe in kayles? Using k numbers we find:

1	1
3	11
5	100
7	10
	122

The sums are not all even, so the position is unsafe in kayles. Only one move by the first player will create a safe pattern, thereby ensuring a win. Can the reader discover it?

NUMBER IN ROW	K-NUMBER	NUMBER IN ROW	K-NUMBER	NUMBER IN ROW	K-NUMBER
1	1	31	10	61	1
2	10	32	1	62	10
3	11	33	1000	63	1000
4	1	34	110	64	1
5	100	35	111	65	100
6	11	36	100	66	111
7	10	37	1	67	10
8	1	38	10	68	1
9	100	39	11	69	1000
10	10	40	1	70	110
11	110	41	100		
12	100	42	111		
13	1	43	10		
14	10	44	1		
15	111	45	1000		
16	1	46	10		
17	100	47	111		
18	11	48	100		
19	10	49	1		
20	1	50	10		
21	100	51	1000		
22	110	52	1		
23	111	53	100		
24	100	54	111		
25	1	55	10		
26	10	56	1		
27	1000	57	100		
28	101	58	10		
29	100	59	111		
30	111	60	100		

NUMBERS OVER 70

REMAINDER	K-NUMBER
0	100
1	1
2	10
3	1000
4	1
5	100
6	111
7	10
8	1
9	1000
10	10
11	111

FIGURE 92
Binary k numbers for playing kayles

The derivation of the k numbers is too complicated to explain here. The interested reader will find it detailed by R. K. Guy and C. A. B. Smith in *Proceedings of the Cambridge Philosophical Society*, Vol. 52, 1956, pages 516–526, and by Thomas H. O'Beirne in *Puzzles and Paradoxes*, Oxford University Press, 1965, pages 165–167. Note that no k number has more than four digits. As a result there are 16 different four-term combinations of odd and even that can occur as column sums, only one of which is even-even-even-even. As Hénon points out, this enables us to conclude, with a high degree of accuracy, that if a kayles starting position is chosen at random from all possible patterns, the probability is close to 1/16 that it will be safe. (The probability approaches 1/16 rapidly as the number of rows increases.)

There are helpful rules that a kayles player can follow without having to analyze each pattern. Two equal chains are safe because whatever your opponent does to one you can do exactly the same to the other. For example, if the two chains are 5 and 5 and he takes the second card in one, you take the second in the other. This leaves chains of 1, 1, 3, 3. If he takes two cards from a 3-chain, you take two from its twin. If he takes a 1-card chain, you take the other. It follows that if the starting position is one single chain, the first player has an easy win. If the chain has one or two cards, he takes them. If it has more than two cards, he takes one or two from the center to leave two equal chains and then continues as explained. If a pattern has an even number of equal pairs of chains, the position is clearly safe, since whatever the first player does to one chain the second player does to that chain's twin.

It is also good to remember the following safe patterns for two or three chains with no more than nine cards in each. The safe doublets (aside from two equal chains, which are always safe) are 1–4, 1–8, 2–7, 3–6, 4–8, and 5–9. Safe triplets can be calculated in the head by memorizing the following three groups: 1, 4, 8; 2, 7; and 3, 6. Any triplet made up of one digit from each group is safe.

Let us turn now to Hénon's two-person string game. We are given an arbitrary number of pieces of string of arbitrary lengths. Players take turns cutting a one-inch segment from any piece of string. The segment can be cut from the end or it can be cut, with two snips, from the interior. In the second case it will leave two pieces of string where there was one before. A one-inch piece can, of course, be taken without any snipping. The person who gets the last one-inch piece wins.

String lengths need not be rational. In Figure 93 a game begins with four strings of lengths 1, pi, the square root of 30, and the square root of 50. Who has the win if both sides play rationally? This seems at first an enormously difficult question, but with the proper insight it is absurdly easy. To work on the problem, rule four straight lines of about the required lengths. As each one-inch segment is erased the remaining lines are labeled with correct lengths.

The game can also be played with closed loops of string. Suppose it begins with seven such loops, each with a length greater than two inches. Without knowing any of the actual lengths, which player has the win? Approached in the right way, this is even easier to answer than the previous question.

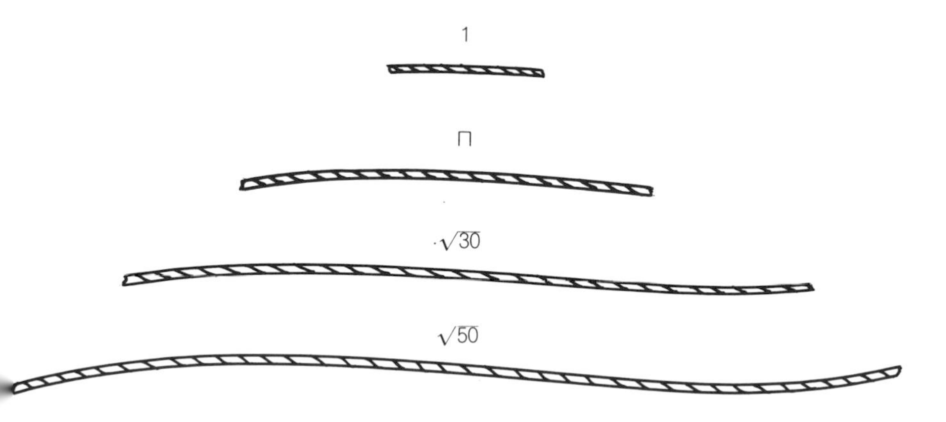

FIGURE 93
Hénon's string game

Our final game is taken from Rufus Isaacs' book *Differential Games* (Wiley, 1965). Devotees of recreational mathematics may recall that Isaacs provided the excellent illustrations for James R. Newman's popular *Mathematics and the Imagination*, but among mathematicians Isaacs is best known as an operations-research expert. His book is filled with original methods of solving difficult conflict games of the kind often encountered in military situations, particularly games that have to do with pursuit and capture. Some of these games are discussed in simpler, discrete versions that have great recreational interest.

One of the book's key games, completely solved by Isaacs, is what he calls the "homicidal-chauffeur game." Imagine a homicidal chauffeur at the wheel of a car he is driving on an infinite plane. He moves at a fixed speed. He can shift the position of his steering wheel instantaneously, but the degree to which he can turn the front wheels is limited. Also on the infinite plane is a lone pedestrian. He can move in any direction at any instant. His speed too is constant, but less than the car's speed. Under what conditions can the car (assumed to be a positive area surrounding the driver) always catch (touch) the pedestrian? Under what conditions can the pedestrian escape permanently? How can the pursuer minimize the time it takes to run down his quarry when this is possible?

Fortunately we shall not be concerned with these difficult questions but with the simpler and somewhat similar game Isaacs calls the "hamstrung squad car." Imagine a city of infinite extent, with streets that form a regular square lattice. A squad car is at one intersection. At another is a carful of criminals. The squad car moves twice as fast as the criminals' car but is hamstrung by having to observe municipal traffic rules that prohibit left turns and U-turns, so that it can only go straight ahead or turn right at each intersection; the criminals' car does not observe these restrictions, so that at each intersection it can move in any of the four directions.

For the quantized game, intersections are replaced by the squares of an infinite checkerboard. The squad car is a counter with a vector arrow painted on it to indicate the direction in which it is moving. The criminals' car is an unmarked counter. Players take turns, the squad car making the first move. All moves are like rook moves in chess: up, down, left, or right but never diagonally. The criminals move one square at a time. The squad car moves two squares, always in a straight line, either in the direction it has been traveling or after making a right turn. (It cannot go one square, turn right and then go another.) It "captures" the criminals if it lands on the square occupied by them or on a square that is adjacent orthogonally or diagonally.

These rules are illustrated in Figure 94. The squad car can move to squares *A* or *B* on its first move. From *A* it can then move to *C* or *D;* from *B,* to *E* or *F*. After each move it should be turned (if necessary) so that its arrow shows the direction in which it was last moving. The criminals can move to squares *W*, *X*, *Y*, or *Z*. If the squad car were on *F* and the criminals were on the same square or on any of the eight shaded squares surrounding it, they would be considered caught.

From what starting positions can the criminals be captured? Isaacs shows that surrounding the squad car's initial square there is an asymmetrical, compact area of exactly 69 squares each of which is a fatal starting spot for the criminals. If they start on any square outside this area they can (assuming an unbounded board) always escape permanently.

The reader is urged to draw a large checkerboard of, say, 50 squares by 50 (or find a room with a suitable floor pattern), choose a starting position near the center for the squad car, and see if he can identify the 69 fatal squares. Until the game is fully analyzed it can provide much amusement. The player moving the criminals' car can choose his starting spot and then see if he can win by reaching the border before he is caught. Playing the game long enough will eventually outline the fatal area, but there is a simpler way by which it can be delineated quickly and each of its squares given a number indicating the number of squad-car moves required for the capture if both sides play rationally.

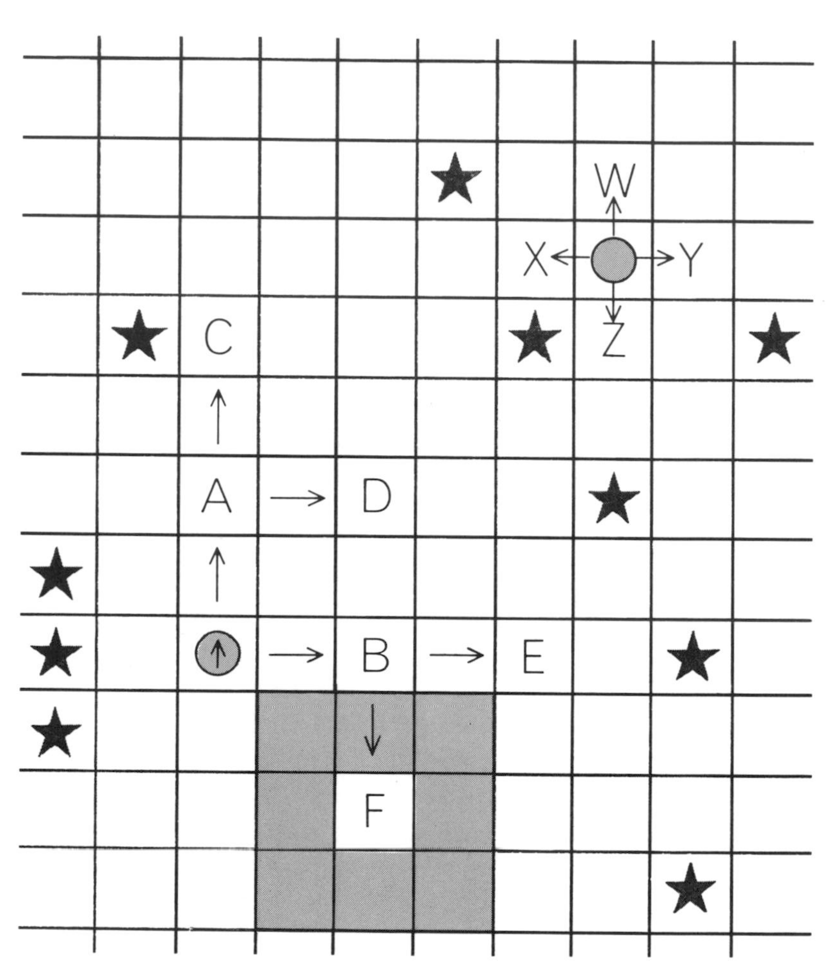

FIGURE 94
Isaacs' "hamstrung squad car" game

For readers not inclined to make a complete analysis here is a simpler problem. Assume that the squad car starts from the position shown in the illustration. The criminals may start from any of the 10 starred squares. From all but one of these starred positions they can escape permanently. Which is the fatal starred square, and in how many moves will the criminals be caught if they start from that square and both sides make their best moves?

ADDENDUM

JOHN HORTON CONWAY reported from Cambridge University that he and some friends had amused themselves by finding sets of nine words, for playing Hot, that could be easily memorized because they could be arranged to make an intelligible sentence. Moreover, the words should be of minimum length: four letters for the word corresponding to the center cell of the ticktacktoe board, three letters for each corner-cell word, and two letters for each side-cell word. Conway came up with: "For stir not so, as if fat ran in," but he thinks this was bettered by Anne Duncan's: "Spit not so, fat fop, as if in pan." The reader should have little trouble arranging them properly on the matrix.

What I called *k* numbers for kayles are usually called Grundy numbers, or Grundy functions, after P. M. Grundy, one of the first to show how such numbers provide strategies for a large family of nim-like games. (See my *Scientific American* column, January, 1972.)

Rufus Isaacs pointed out that a useful generalization of counter take-away games is to imagine a large square lattice with an initial placement of counters, one to a cell, in any desired pattern. Two players alternate in removing any set of counters provided all are in the same row or column. A large family of take-away games can now be viewed as subgames of this game. If the starting position is as shown at the top of Figure 95, we have the *Marienbad* nim game. If the starting position is as shown at the bottom, we have a *Marienbad* game of kayles. If the starting position is a square, and the removed counters must be orthogonally adjacent on each move, we have Piet Hein's tac-tix (see my *Scientific American Book of Mathematical Puzzles & Diversions*, Chapter 25). If only two orthogonally adjacent counters may be removed on each play, we have cram (see my *Scientific American* column, February, 1974). The game can be further generalized, of course, to include other types of lattices, and in *n* dimensions, to create an endless variety of nim-type games.

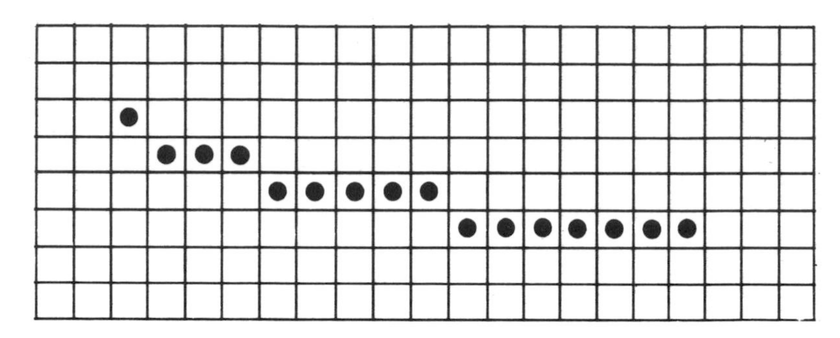

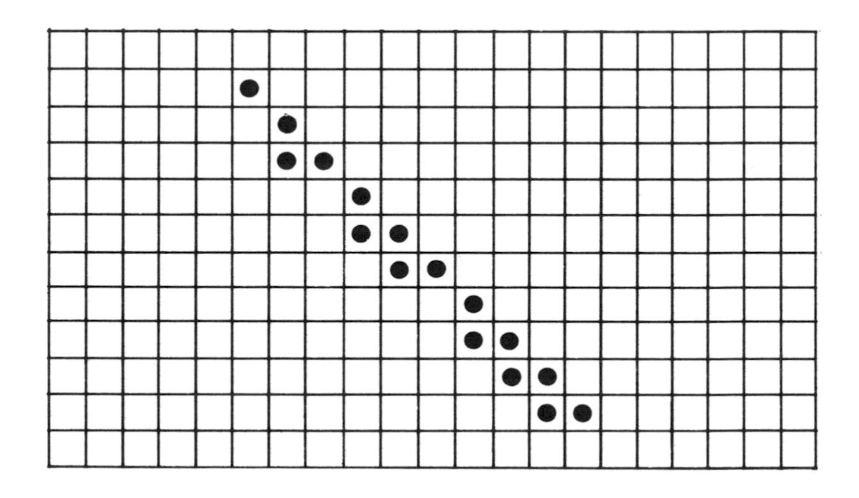

FIGURE 95
Marienbad *versions of nim (above), and kayles (below)*

Several readers called attention to the fact that when the string game is played with a single closed loop it is equivalent to Sam Loyd's daisy game (see *Mathematical Puzzles of Sam Loyd*, Vol. 2, page 40, Dover, 1960).

ANSWERS

READERS WERE asked to search for the winning move of the first player in a game of kayles, starting with the *Marienbad* pattern of four rows of one, three, five, and seven objects. The only winning move is to take the center object from the row of five.

The string game, about which two problems were posed, is isomorphic with kayles! Allowing strings to have irrational lengths seems to complicate the game, but actually it does not, because any fraction over an integral length proves to be irrelevant and can be ignored. Consider a piece of string with a length of 6½ inches. Snipping one inch at a time from one end corresponds to removing one object at a time from the end of a row of six objects in kayles. The leftover fraction—in this case ½ inch—plays no role whatever in the game. Snipping an inch from inside the string, starting, say, ¾ inch (or any fraction between ½ and 1) from the end, corresponds to removing two objects from the end of a row of six in kayles; the ¾-inch piece obviously plays no further role in the game, and you are left with a piece of 4¾ inches, which is equivalent to a row of four in kayles. Snipping an inch from the interior of the 6½-inch piece, an integral number of inches from one end, corresponds to removing one object from the interior of a row of six objects in kayles. Snipping an interior inch that is an integral number of inches, plus a fraction between ½ and 1, from the end of the 6½-inch piece corresponds to removing two adjacent objects from inside a row of six in kayles.

A little reflection and testing will soon convince you that every move in kayles has its counterpart in the string game, and vice versa. Each piece of string corresponds to a row in kayles with a number of objects equal to the number of whole inches in the string.

Once the equivalence of the two games is recognized, the first question about the string game is immediately answered. If strings have lengths of 1, pi (3.14+), the square root of 30 (5.47+), and the square root of 50 (7.07+), the strings are equivalent to the *Marienbad* pattern of rows with one, three, five, and seven objects. The first player can win, therefore, as explained: he snips an inch that is two inches from one end of the 5.47+ string, then continues with moves that correspond to the kayles strategy as outlined earlier.

If the string game is played with any number of closed loops, each longer than two inches, the second player has an easy win. Whenever his opponent opens a loop by removing an inch from it, the second player simply removes an inch from the exact middle of the same string. This leaves two equal pieces. As in kayles, this is a safe pattern, for whatever his opponent does to one piece, the second player does to the other piece. Thus the pattern quickly becomes a set of pairs of duplicate strings, and therefore the second player is sure to get the last inch.

If the starting pattern includes one closed loop at least an inch long but no more than two inches long, the first player wins by taking an inch from it and then playing the strategy just described with the remaining pieces. It is easy to see that the first player wins if there is an odd number of such small loops and loses if there is an even number of loops.

Figure 96 shows the area of fatal starting positions for the criminals' car in the "hamstrung squad car" game. The squad car starts at the position shown by the counter with the vector arrow. The criminals' car can be captured if it starts on any numbered square. The number on each square is the number of squad-car moves required for the capture if both sides play rationally. The same illustration also answers the final question: Of the 10 starred starting positions that were shown for the criminals, the only fatal square is the one (gray) that is a knight's move left and down from the squad car. Nine squad-car moves effect the capture if both sides make their best moves.

For a simple procedure by which the positions and numbers of moves can be obtained the reader is referred to Rufus Isaacs' *Differential Games*, pages 56–62, in which the game is analyzed and from which the illustration of the answer is taken. I leave it to the reader to work out the strategies by which the squad car

			8	11					
			5	8					
		3	4	7	10				
		2	3	4	7				
	1	1	1	3	6	9			
	1	1	1	2	3	6			
	0	0	0	1	1	5	8		
	0		0	1	1	2	3		
9	0	0	0	1	1	3	4	5	8
	7	4	3	2	3	4	7	8	11
	10	7	4	3	6	7	10		
		8	5	6	9				
		11	8						

FIGURE 96

Solution to the "hamstrung squad car" game

captures the criminals' car with the minimum number of moves and by which the criminals either delay their capture as long as possible or escape permanently if they start from an unnumbered square or if the police blunder.

CHAPTER 17

Cooks and Quibble-Cooks

Too many cooks spoil the broth.
—OLD ENGLISH PROVERB

WHEN A mathematical puzzle is found to contain a major flaw—when the answer is wrong, when there is no answer, or when, contrary to claims, there is more than one answer or a better answer—the puzzle is said to be "cooked." The expression was taken over from the argot of chess. (*The Oxford English Dictionary* quotes an 1899 statement about chess problems to the effect that "if there are two key-moves, a problem is cooked.")

An entire book could be written about the more amusing instances of chess problems and game analyses by experts that were later cooked by other experts. In *Curious Chess Facts* (Black Knight Press, 1937) Irving Chernev cites what is surely one of the most embarrassing chess mistakes ever to appear in print. The eighth edition of a popular late-19th-century German handbook on chess openings, by Jean Dufresne and Jacques Mieses, gave the following line of play for a queen's gambit declined. (In this notation *N* is the symbol for knight.)

WHITE	BLACK
1. P—Q4	P—Q4
2. P—QB4	P—K3
3. N—QB3	P—QB4
4. N—B3	BP × P
5. KN × P	P—K4
6. KN—N5	P—Q5
7. N—Q5	N—QR3
8. Q—R4	B—Q2
9. P—K3	N—K2

Black, the authors wrote, now has a "superior position." The fact is, however, that White can checkmate on his next move. Readers may enjoy playing to the position and seeing how quickly they can spot the mating play.

Tournament chess games between grandmasters are often won because one of the players has managed to cook a standard line of opening play, but kept the cook to himself until he could use it against a worthy opponent. Checkers has been so exhaustively analyzed that most games between top experts are draws. When a win occurs, it is usually the result of an expert springing a secret, unpublished cook on a familiar line of play.

Science progresses, of course, by a never-ending series of cooks. Indeed, as the philosopher Karl Popper has emphasized, a scientific theory is "empty" if there is no conceivable way to cook it; the more ways there are in which a theory might be cooked, the stronger the theory is if it ultimately passes all the tests. Mathematics is regarded as having an iron certainty not possessed by science, but mathematicians can make mistakes and so even in mathematics a proof has to be established by the social process of confirmation by others. The history of mathematics is filled with "proofs" by eminent mathematicians that were later cooked. This is particularly true of recreational mathematics, a field dominated by amateurs.

Sam Loyd, the greatest of American puzzle inventors, published such a vast quantity of chess problems and mathematical puzzles that it is not surprising scores of his ingenious creations turned out to have fatal flaws. One of his worst mistakes was his solution to a dissection problem that is reprinted on page 27 of his *Cyclopedia of Puzzles*. The reader is asked to cut the figure shown at the left in Figure 97—a square missing a quarter-section—into the fewest pieces that can be rearranged to make a perfect square. Loyd's four-piece answer is shown by the broken lines in the figure at the left and the reassembled pieces are shown at the right. "There are numerous ways of performing the feat with from five to a dozen pieces." Loyd wrote, "but the answer given is both difficult and scientific."

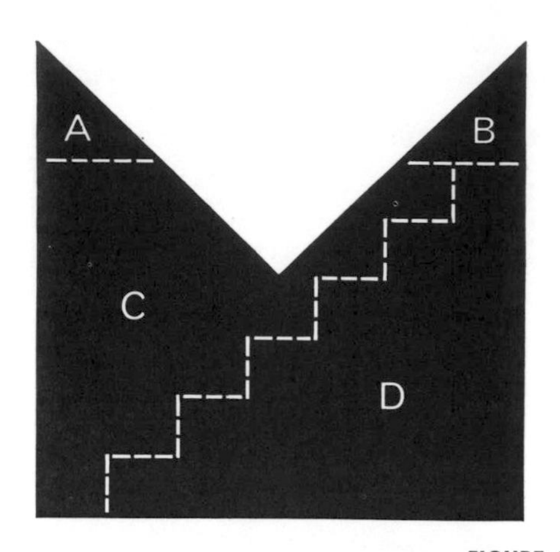

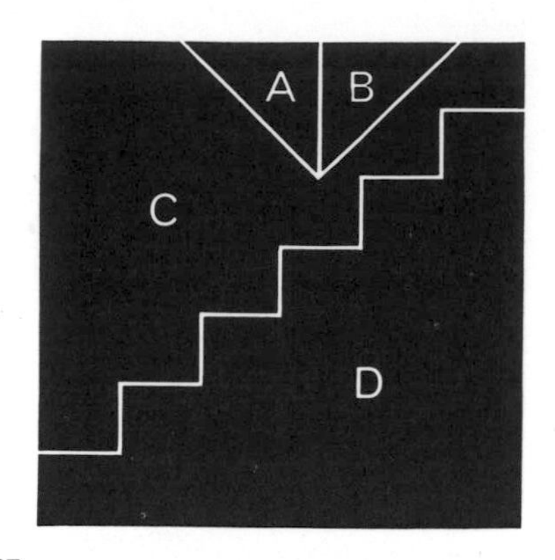

FIGURE 97

A Loyd miter-to-square dissection that was cooked by Dudeney

It was the British puzzle expert Henry Ernest Dudeney, a better mathematician than Loyd, who cooked this puzzle. Packing the little triangles into the "valley" forms a rectangle, which is then presumably converted by a "stair-step" principle into a square. But the step principle does the trick only if the sides of the rectangle are in certain ratios, and the ratio in this case (three to four) is not one of them. (See Dudeney, *Amusements in Mathematics*, Problem 150, and *Modern Puzzles*, Problem 115.) Loyd's clever dissection produces not a square but an oblong. Dudeney presented a correct five-piece dissection [*see Figure 98*]. No solution with four pieces is believed to be possible, but Harry Lindgren, in his beautiful book *Recreational Problems in Geometric Dissections* (Dover, 1972), shows how two of the miter-shaped pieces can each be sliced in the same way into four parts and the eight pieces re-formed to make two congruent squares [*see Figure 99*].

Sometimes a puzzle is cooked and then the cook is cooked. Angelo Lewis, an Englishman who wrote books on magic and puzzles under the pseudonym "Professor Hoffmann," gave a puzzle with 20 counters in his book *Puzzles Old and New* (1893): In the formation in Figure 100, how many different squares are indicated by four counters at their corners? Seventeen, Hoffmann said. In an article "The Best Puzzles with Coins" (*The Strand Magazine*, 1909) Dudeney cooked this

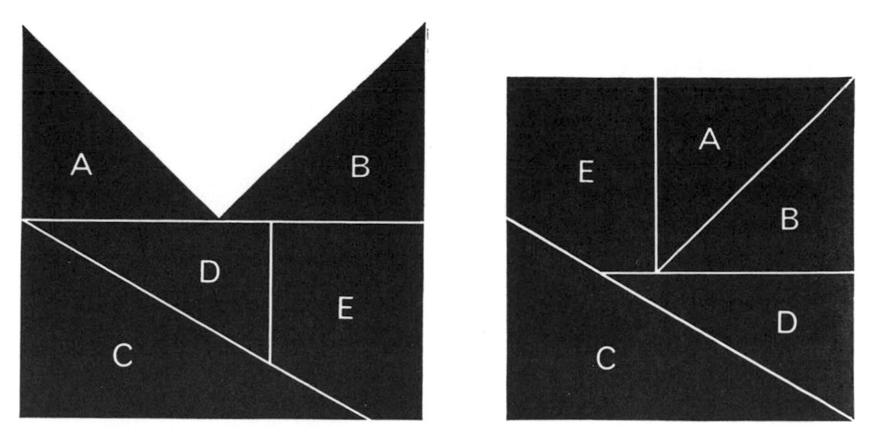

FIGURE 98
Dudeney's correct five-piece dissection

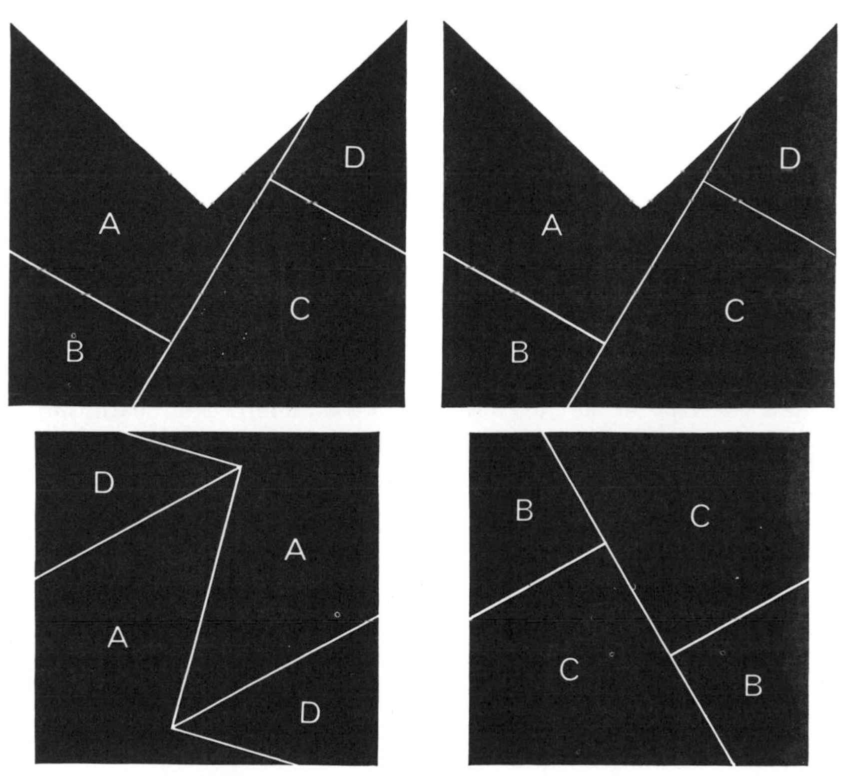

FIGURE 99
Lindgren's dissection of two miters to two squares

statement by listing 19 different squares. In actuality there are 21; Dudeney gave the correct figure when he reprinted the puzzle in one of his books. The reader should have no difficulty finding the 21 squares, but the second part of this old puzzle is less easy: Remove six counters so that no square of any size remains indicated by four counters at the corners.

FIGURE 100
A twice-cooked counter puzzle

Most of Dudeney's errors were caught by readers of his magazine and newspaper columns, enabling him to correct them before the puzzles appeared in books. But even his books contain many cookable puzzles. Consider the following problem of the rook's-tour type, which appears in *Amusements in Mathematics* (Problem 244) and *Modern Puzzles* (Problem 161). A car starts at intersection *A* at the edge of a square city area seven blocks on the side [*see Figure 101*]. Alternatively, one could place a rook on the king's square of a standard chessboard, but moving along lattice lines makes the question of distance less ambiguous. The car must travel the longest route possible with-

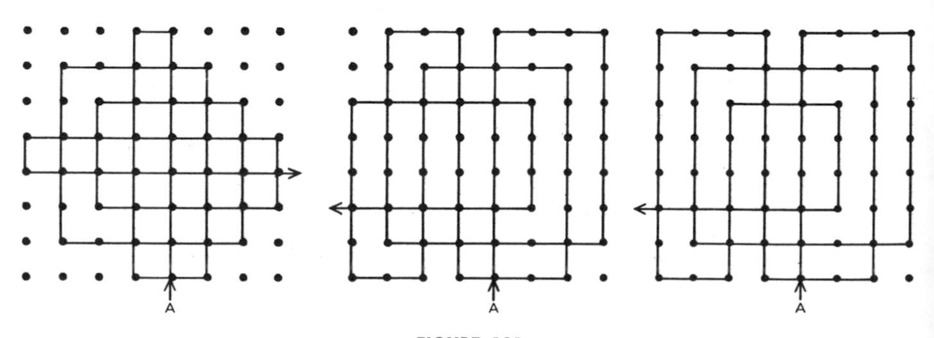

FIGURE 101
First solution to a graph puzzle (left) and two cooks

out making more than 15 turns and without going over any part of its path twice. In achieving the maximum distance it must also leave the fewest possible intersections unvisited.

An inferior solution [*at left in Figure 101*] was given in the two Dudeney books: a path of 70 blocks with 19 lattice points unvisited. Dudeney himself cooked this with the path shown in the middle picture (solution to Problem 269 in his later book, *Puzzles and Curious Problems*), which is 76 blocks long and leaves only three intersections untouched. Is this the ultimate answer? No; Victor Meally of Dublin County in Ireland sent me the path shown at the right in the illustration: 76 blocks, 15 turns, and only *one* corner unvisited! Is it possible to cook the problem again by finding a 15-turn path longer than 76 blocks, or a 76-block path that visits *all* intersections? Probably not, but as far as I know Meally's solution has not been proved final.

Problem 57 in *Puzzles and Curious Problems* shows a clock-face with Roman numerals and asks how it can be cut into four pieces each bearing numerals that add up to 20. Since the numbers 1 through 12 add up to 78, some device must be found for raising the total to 80. Dudeney's clumsy method was to view the IX upside down as XI, making possible the dissection shown at the left in Figure 102. Loyd removed this blemish by publish-

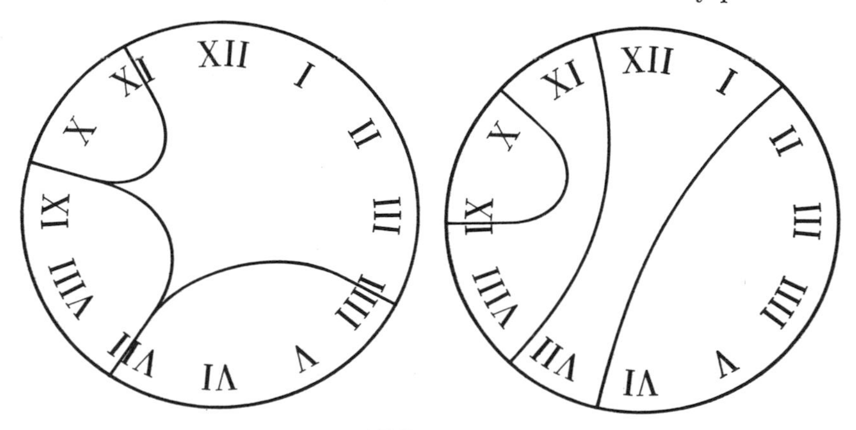

FIGURE 102
Dudeney's clock-puzzle answer (left) and Loyd's cook (right)

ing (in 1909, he reports in his book *Sam Loyd and His Puzzles*) the dissection shown at the right in the illustration. Loyd in turn overlooked a dozen other equally perfect solutions, however, none of which requires that a numeral be viewed from the wrong side. The reader should be able to find nine of them without much effort, but three are quite elusive. Note that the Roman numeral for 4 is written IIII, according to the usual custom among clockmakers, rather than IV. The numerals must be regarded as being permanently attached to the rim of the clockface; that is, a dissection line may go *through* one of the hours, but it is not allowed to loop *around* any numerals, separating them from the rim. If this were allowed, the problem would lose interest, because hundreds of solutions would be possible.

In editing Loyd's mammoth *Cyclopedia* for two Dover paperback collections, I found hundreds of mistakes, most of them printer's errors. Among several legitimate cooks that I missed, one of the most confusing (the confusions are of the kind that must constantly plague those who keep records on artificial satellites) has to do with Loyd's eagle problem on page 117 of the *Cyclopedia* and was first called to my attention by D. H. Wheeler of Minneapolis. Exactly at sunrise an American eagle takes off from the top of the Capitol dome in Washington, flies due east until the sun is overhead, and then reverses its direction and flies due west until it sees the sun set. Since the eagle and the sun move in opposite directions during the morning part of the flight and in the same direction during the afternoon part, it is clear that the afternoon flight will be longer and that the eagle will end it at a spot west of where it started. The eagle rests until sunrise and repeats the sequence—it flies east until it sees the sun at noon, flies west until it sees the sun set, and then rests until the next sunrise—until it eventually works its way around the earth to Washington again. Assume that the circumference of the flight circle, from the dome around the earth on an east-west path and back to the dome, is exactly 19,500 miles. Assume also that at the end of each "day," as observed by the eagle, it ends its flight 500 miles west of where it started at sunrise. When the eagle gets back to the Capitol, how many "days" have elapsed as measured by someone in Washington? The answer 38 days, given in the first of the two paperback collections, is wrong. How does the reader calculate it?

Geography is also involved in one of the most brilliant of all puzzle cooks. An explorer stands at a certain spot on the earth. Looking due south, he sees a bear 100 yards away. The bear walks 100 yards due east while the explorer stands still. The explorer then points his gun due south, fires and kills the bear. Where is the man standing? The original answer was, of course: at the North Pole. As explained in *The Scientific American Book of Mathematical Puzzles & Diversions* (Simon & Schuster, 1959), however, the problem has another answer. The man could be standing very close to the South Pole—so close that when the bear walks east, the 100-yard path carries it once around the pole and back to where it started. Actually there is an infinite set of answers of this type, because the man could stand still closer to the South Pole, allowing the bear to circle the pole twice, or three times, and so on. Is the problem now completely cooked? Far from it. Benjamin L. Schwartz, writing in a mathematics journal six years ago, found two more completely different families of solutions! Read the problem again carefully and see if you can think of them.

In addition to genuine cooks of problems in my *Scientific American* columns, I also sometimes receive from astute readers what, for want of a better name, I call the quibble-cook. This is a cook that takes advantage of a play on words, or a lack of precision in stating a problem's conditions.

I once gave this joke problem in a children's book. Circle six digits in the following table so that the total of circled numbers is 21:

9 9 9
5 5 5
3 3 3
1 1 1

My answer was to turn the page upside down, circle each of the three 6's and each of the three 1's. Howard R. Wilkerson of Silver Spring, Md., delightfully quibble-cooked this problem by finding a much better solution. Without inverting the page he circled each of the 3's, circled the 1 on the left and then drew one larger circle around the other two 1's. The sum of the circled numbers, 3, 3, 3, 1, 11, obviously is 21.

ADDENDUM

NO READERS found a 15-turn path for the rook tour longer than 76 blocks, or a 76-block path that visited all intersections. Many readers sent 75-block paths with 15 turns and no intersections missed. It has long been established, by the way, that 14 turns is minimum for a rook tour touching all cells, and 15 if the tour is reentrant.

The problem about shooting the bear brought a number of letters suggesting other solutions: the explorer looks south into a mirror, the explorer "stands still" on a moving vehicle or boat, the bear stays in the same spot while "walking" on a moving ice floe, the bullet completely circles the earth, and so on. The following letter from R. S. Burton, of Shepperton, England, was printed in the Letters department of *Scientific American*, October, 1966:

SIRS:

In the interests of bear-shooting I wish to submit another infinite set of answers to the problem. . . . My solution alleviates the frustrations attendant on this form of hunting in antarctic regions due to the scarcity of the genus Ursus *at these latitudes. My proposal is such that, although the bear must be in the Southern Hemisphere, the explorer can position himself anywhere on the earth's surface at a similar longitude to the long-suffering beast and shoot it at any range whatever to the south.*

The method is based on the fact that an easterly component of velocity is imparted to a bullet fired from a gun pointed in a southerly direction, owing to the earth's spin. This component is equal to the peripheral speed of the earth at the Equator times the sine of the latitude of the explorer. Providing that the bear is at a greater angle of latitude (south) than the explorer (north or south) and the latter's weapon is of sufficiently long range and/or low muzzle velocity, the bear will be shot. As an example, if the explorer is at latitude 89 degrees south and the bear at 89 degrees 10 minutes south (that is, about 11½ miles distant), a bullet fired due south at about 600 miles per hour will still hit the bear after it has moved 100 yards to the east. Smaller ranges and/or latitudes at which the encounter between explorer and bear takes place will necessitate a weapon of lower muzzle velocity and vice versa.

It is of interest in this connection that during World War I the Germans had to make allowance for the difference in latitude between their long-range "Paris guns" and the target, aiming off to the east in order to counteract this effect of the earth's rotation.

Benjamin Schwartz's unpublished reply to the above letter follows:

DEAR MR. BURTON:

I appreciate your addition to the literature of bear-hunting lore. But I respectfully suggest that you are actually changing the rules of the problem in your solution. In my 1960 paper, I was quite explicit in making the problem an exercise in spherical mathematics. It is couched in the language of bear-hunting for interest, but it is fundamentally an exercise in coordinate geometry. Note particularly the marked passages in the enclosed reprint.

For your version to be valid, the domain of discussion must be expanded to include dynamics. And once you do this, you are subject to new attack. For example, the low muzzle velocities you speak of preclude attaining the range needed. Gravity, you know. For the case given in your letter, the shot would have had to have been fired at an altitude of about 75 thousand feet to avoid hitting the ground before it goes 11½ miles horizontally (neglecting air resistance). Now, that presupposes a pretty tall hunter, I think you have to agree.

One way out of this difficulty is to assume the hunter fires up at a steep angle, so that the horizontal component of velocity is 600 mph, as given by you; and the vertical component is chosen to have the bullet return to earth at just the right moment, 69 seconds later. I fear however that if we propose this, the next cook will demand that air resistance and exterior ballistics be considered. You see the hornet's nest you've opened.

To retain the Coriolis effect while keeping the problem "theoretical," what do you think of the following "solution"? The shot has a muzzle velocity of about 17,000 mph, just sufficient to keep the bullet in orbit at an altitude of 5 feet above the ground. It continues to circle the earth indefinitely with its orbit precessing to the west until (with probability 1) it hits any bear of height more than 5 feet, and any positive width. (Note: after shooting, the hunter is required to duck.)

ANSWERS

1. WHITE CHECKMATES with knight to Q6.

2. Figure 103 shows how to remove six (gray) of the 20 counters so that no four remaining counters mark the corners of any of the 21 squares. The solution is unique except for rotations and reflections.

3. Apart from Sam Loyd's solution to the clock puzzle, there are 12 other perfect answers [*see Figure 104*]. Each clockface is divided into four parts, the numbers in each part adding up to 20. The last three solutions are the hardest to find.

4. Loyd's eagle completes its flight at sundown after 39½ days as measured in Washington. The eagle will have seen 38½

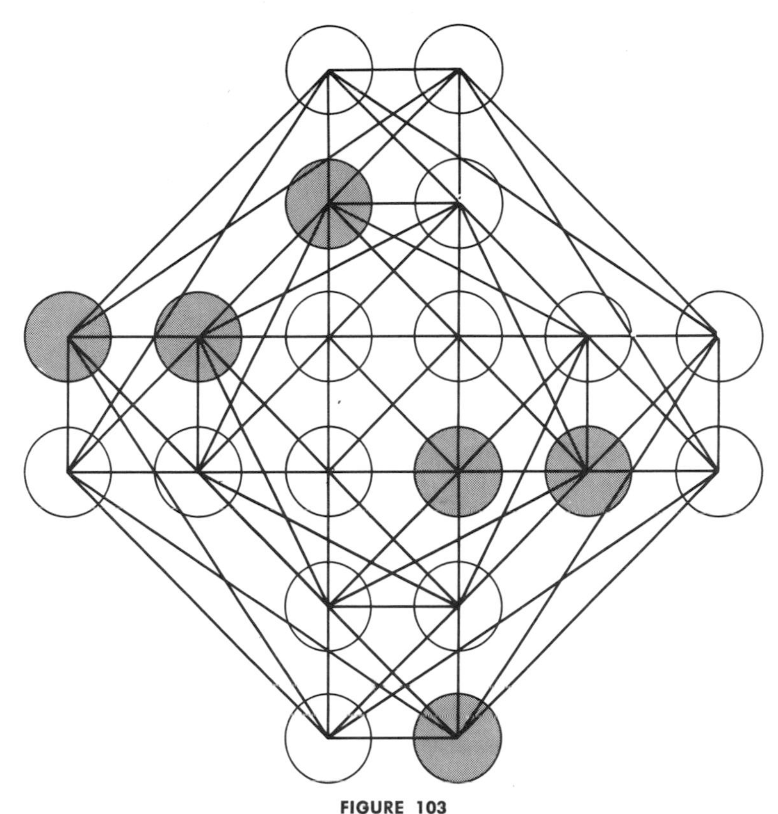

FIGURE 103
Solution to 20-counter problem

"days" (measured by the sunrises and sunsets it saw in flight) but, because it circled the earth in a direction opposite to the earth's spin, it has lost a day compared with the elapsed days in Washington.

5. Assume that the explorer and bear are near the South Pole. The bear is 100 yards south of the man, at a spot such that, when it completes its 100-yard walk east, it will be exactly opposite the man on the *other side* of the South Pole. Thus when the explorer aims his gun south and fires, the bullet travels over the South Pole and hits the bear. There is an infinite set of solutions, for the bear can be closer still to the pole so that his walk takes him one and a half times around the pole, or two and a half times, and so on.

FIGURE 104
Solutions to clock-dissection problem

The second easily overlooked family of answers hinges on the statement that the man "looking due south . . . sees a bear 100 yards away." Clearly the man and bear can have initial positions in which they are on opposite sides of the South Pole but 100 yards apart. The man is farther from the pole than the bear is. After the bear has walked 100 yards due east it is halfway around a circle to a spot directly south of the man, on the *same* side of the pole. Of course, the man can be a bit farther from the pole, so that the bear makes a complete circle, or one and a half circles, or two circles, or two and a half circles, and so on, to generate another infinite family of solutions that reaches a limit when the man is 100 yards from the pole. The bear's walk then degenerates into a pirouette on the pole itself. Both families of overlooked solutions are given by Benjamin Schwartz in his article "What Color Was the Bear?" in *Mathematics Magazine*, Vol. 34, September–October, 1960, pages 1–4.

CHAPTER 18

Piet Hein's Superellipse

There is
one art,
no more,
no less:
to do
all things
with art-
lessness.
—PIET HEIN

CIVILIZED MAN is surrounded on all sides, indoors and out, by a subtle, seldom-noticed conflict between two ancient ways of shaping things: the orthogonal and the round. Cars on circular wheels, guided by hands on circular steering wheels, move along streets that intersect like the lines of a rectangular lattice. Buildings and houses are made up mostly of right angles, relieved occasionally by circular domes and windows. At rectangular or circular tables, with rectangular napkins on our laps, we eat from circular plates and drink from glasses with circular cross sections. We light cylindrical cigarettes with matches torn from rectangular packs, and we pay the rectangular bill with rectangular bank notes and circular coins.

Even our games combine the orthogonal and the round. Most outdoor sports are played with spherical balls on rectangular fields. Indoor games, from pool to checkers are similar combinations of the round and the rectangular. Rectangular playing cards are held in a fanlike circular array. The very letters on this rectangular page are patchworks of right angles and circular arcs. Wherever one looks the scene swarms with squares and circles and their affinely stretched forms: rectangles and ellipses. (In a sense the ellipse is more common than the circle, because every circle appears elliptical when seen from an angle.) In Op paintings and textile designs, squares, circles, rectangles, and ellipses jangle against one another as violently as they do in daily life.

The Danish writer and inventor Piet Hein recently asked himself a fascinating question: What is the simplest and most pleasing closed curve that mediates fairly between these two clashing tendencies? Originally a scientist, Piet Hein (he is always spoken of by both names) is well known throughout Scandinavia and English-speaking countries for his enormously popular volumes of gracefully aphoristic poems (which critics have likened to the epigrams of Martial) and for his writings on scientific and humanistic topics. To recreational mathematicians he is best known as the inventor of the game Hex, of the Soma cube, and of other remarkable games and puzzles. He was a friend of Norbert Wiener, whose last book, *God and Golem, Inc.*, is dedicated to him.

The question Piet Hein asked himself had been suggested by a knotty city-planning problem that first arose in 1959 in Sweden. Many years earlier Stockholm had decided to raze and rebuild a congested section of old houses and narrow streets in the heart of the city, and after World War II this enormous and costly program got under way. Two broad new traffic arteries running north-south and east-west were cut through the center of the city. At the intersection of these avenues a large rectangular space (now called Sergel's Square) was laid out. At its center is an oval basin with a fountain surrounded by an oval pool containing several hundred smaller fountains. Daylight filters through the pool's translucent bottom into an oval self-service restaurant, below street level, surrounded by oval rings of pillars and shops. Below that there eventually will be two more oval floors for dining and dancing, cloakrooms, and kitchen.

In planning the exact shape of this center the Swedish architects ran into unexpected snags. The ellipse had to be rejected because its pointed ends would interfere with smooth traffic flow around it; moreover, it did not fit harmoniously into the rectangular space. The city planners next tried a curve made up of eight circular arcs, but it had a patched-together look with ugly "jumps" of curvature in eight places. In addition, plans called for nesting different sizes of the oval shape, and the eight-arc curve refused to nest in a pleasing way.

At this stage the architectural team in charge of the project consulted Piet Hein. It was just the kind of problem that appealed to his combined mathematical and artistic imagination, his sense of humor, and his knack of thinking creatively in unexpected directions. What kind of curve, less pointed than the ellipse, could he discover that would nest pleasingly and fit harmoniously into the rectangular open space at the heart of Stockholm?

To understand Piet Hein's novel answer we must first consider the ellipse, as he did, as a special case of a more general family of curves with the following formula in Cartesian coordinates,

$$\left|\frac{x}{a}\right|^n + \left|\frac{y}{b}\right|^n = 1,$$

where a and b are unequal parameters (arbitrary constants) that represent the two semiaxes of the curve, and n is any positive real number. The vertical brackets indicate that each fraction is to be taken with respect to its absolute value; that is, its value without regard to sign. (Brackets will be omitted in some formulas to be given later; assume that absolute values are intended.)

When $n = 2$, the real values of x and y that satisfy the equation (in modern jargon, its "solution set") determine the points on the graph that lie on an ellipse with its center at the origin of the two coordinates. As n decreases from 2 to 1, the oval becomes more pointed at its ends ("subellipses," Piet Hein calls them). When $n = 1$, the figure is a parallelogram. When n is less than 1, the four sides are concave curves that become increasingly concave as n approaches 0. At $n = 0$ they degenerate into two crossed straight lines.

If n is allowed to increase above 2, the oval develops flatter and flatter sides, becoming more and more like a rectangle; indeed, the rectangle is its limit as n approaches infinity. At what point is such a curve most pleasing to the eye? Piet Hein settled on $n = 2\frac{1}{2}$. With the help of a computer, 400 coordinate pairs were calculated to 15 decimal places, and larger, precise curves were drawn in many different sizes, all with the same height-width ratios (to conform with the proportions of the open space at the center of Stockholm). The curves proved to be strangely satisfying, neither too rounded nor too orthogonal, a happy blend of elliptical and rectangular beauty. Moreover, such curves could be nested, as shown in Figures 105 and 106, to give a strong feeling of harmony and parallelism between the concentric ovals. Piet Hein calls all such curves with exponents above 2 "superellipses." Stockholm immediately accepted the $2\frac{1}{2}$-exponent superellipse as the basic motif of its new center. When the entire center is finally completed it will surely be one of the great tourist attractions (certainly for mathematicians!) of Sweden. Already the large superelliptical pool has conferred upon Stockholm an unusual mathematical flavor, like the big catenary curve of St. Louis's Gateway Arch, which dominates the local skyline.

Meanwhile Piet Hein's superellipse has been enthusiastically adopted by Bruno Mathsson, a well-known Swedish furniture designer. He first produced a variety of superelliptical desks, now in the offices of many Swedish executives, and has since followed with superelliptical tables, chairs, and beds. (Who needs the corners?) Industries in Denmark, Sweden, Norway, and Finland have turned to Piet Hein for solutions to various orthogonal-versus-circular problems, and in recent years he

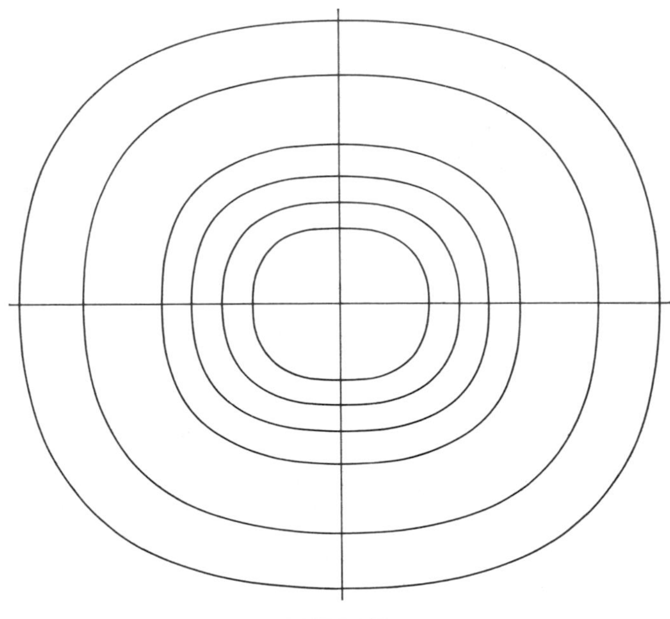

FIGURE 105
Concentric superellipses

has been working on superelliptical furniture, dishes, coasters, lamps, silverware, textile patterns, and so on. The tables, chairs, and beds embody another Piet Hein invention: unusual self-clamping legs that can be removed and attached with great ease.

"The superellipse has the same convincing unity as the circle and ellipse, but it is less obvious and less banal," Piet Hein wrote recently in the leading Danish magazine devoted to applied arts and industrial design. (The magazine's white cover for that issue bore only the stark black line of a superellipse, captioned with the formula of the curve.)

"The superellipse is more than just a new fad," Piet Hein continued; "it is a relief from the straitjacket of the simpler curves of first and second powers, the straight line and the conic sections." Incidentally, one must not confuse the Piet Hein superellipse with the superficially similar potato-shaped curves one often sees, particularly on the face of television sets. These are seldom more than oval patchworks of different kinds of arc,

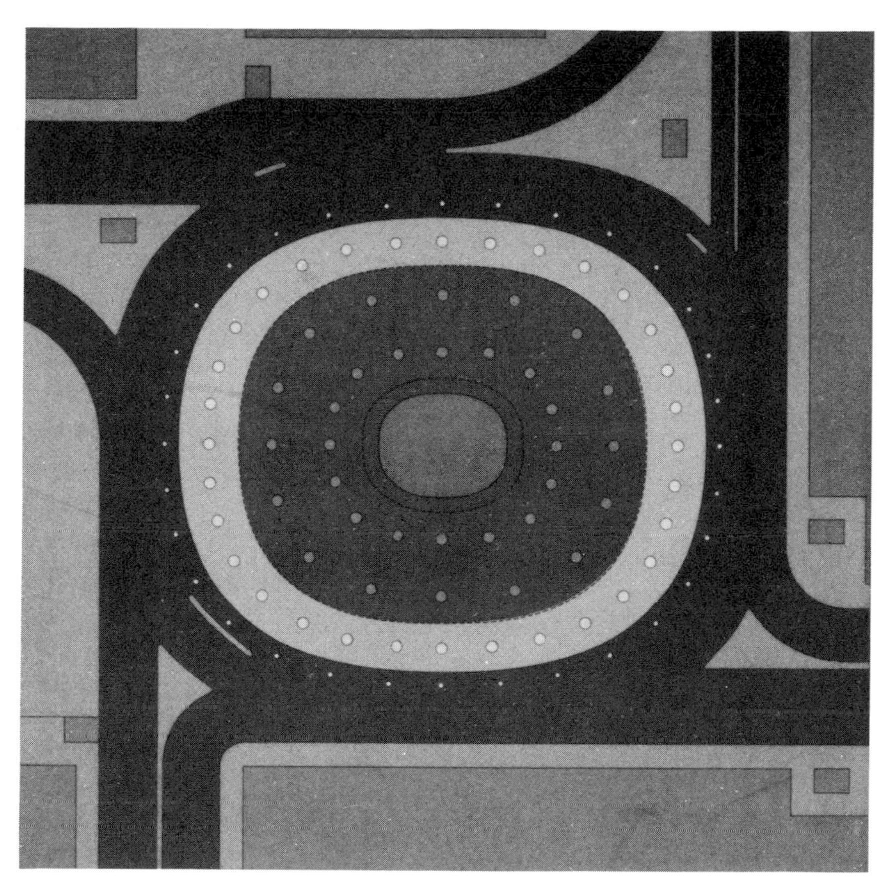

FIGURE 106

Plan of Stockholm's underground restaurants and the pools above them

and they lack any simple formula that gives aesthetic unity to the curve.

When the axes of an ellipse are equal, it is of course a circle. If in the circle's formula, $x^2 + y^2 = 1$, the exponent 2 is replaced by a higher number, the graphed curve becomes what Piet Hein calls the "supercircle." At $2\frac{1}{2}$ it is a genuine "squared circle" in the sense that it is artistically midway between the two extremes. The changing shapes of curves with the general formula $x^n + y^n = 1$, as n varies from 0 to infinity, are graphed in Figure 107. If the graph could be stretched uniformly along one axis (one of the affine transformations), it would depict the family of curves of which the ellipse, subellipses, and superellipses are members.

In the same way, one can raise the exponent in the corresponding Cartesian formulas for spheres and ellipsoids to obtain what Piet Hein calls "superspheres" and "superellipsoids." If the exponent is $2\frac{1}{2}$, such solids can be regarded as spheres and ellipsoids that are halfway along the road to being cubes and bricks.

The true ellipsoid, with three unequal axes, has the formula

$$\frac{x^2}{a^2} + \frac{y^2}{b^2} + \frac{z^2}{c^2} = 1,$$

where a, b, and c are unequal parameters representing half the length of each axis. When the three parameters are equal, the figure is a sphere. When only two are equal, the surface is called an "ellipsoid of rotation" or a spheroid. It is produced by rotating an ellipse on either of its axes. If the rotation is on the longer axis, the result is a prolate spheroid—a kind of egg shape with circular cross sections perpendicular to the axis.

It turns out that a solid model of a prolate spheroid, with homogeneous density, will no more balance upright on either end than a chicken egg will, unless one applies to the egg a stratagem usually credited to Columbus. Columbus returned to Spain in 1493 after having discovered America, thinking that the new land was India and that he had proved the earth to be round. At Barcelona a banquet was given in his honor. This is how Girolamo Benzoni, in his *History of the New World* (Venice, 1565), tells the story (I quote from an early English translation):

Columbus, being at a party with many noble Spaniards . . . one of them undertook to say: "Mr. Christopher, even if you had not found the Indies, we should not have been devoid of a man who would have attempted the same thing that you did, here in our own country of Spain, as it is full of great men clever in cosmography and literature." Columbus said nothing in answer to these words, but having desired an egg to be brought to him, he placed it on the table saying: "Gentlemen, I will lay a wager with any of you, that you will not make this egg stand up as I will, naked and without anything at all." They all tried, and no one succeeded in making it stand up. When the egg came round to the hands of Columbus, by beating it down on the table he fixed it, having thus crushed a little of one end; wherefore all remained confused, understanding what he would have said: That after the deed is done, everybody knows how to do it.

The story may be true, but a suspiciously similar story had been told 15 years earlier by Giorgio Vasari in his celebrated *Lives of the Most Eminent Painters, Sculptors and Architects* (Florence, 1550). Young Filippo Brunelleschi, the Italian architect, had designed an unusually large and heavy dome for Santa Maria del Fiore, the cathedral of Florence. City officials had asked to see his model, but he refused, "proposing instead . . . that whosoever could make an egg stand upright on a flat piece of marble should build the cupola, since thus each man's intellect would be discerned. Taking an egg, therefore, all those Masters sought to make it stand upright, but not one could find a way. Whereupon Filippo, being told to make it stand, took it graciously, and, giving one end of it a blow on the flat piece of marble, made it stand upright. The craftsmen protested that they could have done the same; but Filippo answered, laughing, that they could also have raised the cupola, if they had seen the model or the design. And so it was resolved that he should be commissioned to carry out this work."

The story has a topper. When the great dome was finally completed (many years later, but decades before Columbus' first voyage), it had the shape of half an egg, flattened at the end.

What does all this have to do with supereggs? Well, Piet Hein (my source, by the way, for the references on Columbus and

Brunelleschi) discovered that a solid model of a $2\frac{1}{2}$-exponent superegg—indeed, a superegg of *any* exponent—if not too tall for its width, balances immediately on either end without any sort of skulduggery! Indeed, dozens of chubby wooden and silver supereggs are now standing politely and permanently on their ends all over Scandinavia.

Consider the wooden superegg shown in Figure 108, which has an exponent of $2\frac{1}{2}$ and a height-width ratio of 4 : 3. It *looks* as if it should topple over, but it does not. This spooky stability of the superegg (on both ends) can be taken as symbolic of the superelliptical balance between the orthogonal and the round, which is in turn a pleasant symbol for the balanced mind of individuals such as Piet Hein who mediate so successfully between C. P. Snow's "two cultures."

FIGURE 108
Wooden superegg, stable on either end

ADDENDUM

THE FAMILY of plane curves expressed by the formula $|x/a|^n + |y/b|^n = 1$ was first recognized and studied by Gabriel Lamé, a 19th-century French physicist, who wrote about them in 1818. In France they are called *courbes de Lamé;* in Germany, *Lamesche kurven.* The curves are algebraic when n is rational, transcendent when n is irrational.

When $n = 2/3$ and $a = b$ [*see Figure 107*] the curve is an astroid. This is the curve generated by a point on a circle that is one-fourth or three-fourths the radius of a larger circle, when the smaller circle is rolled around the inside of the larger one. Solomon W. Golomb called attention to the fact that if n is odd, and the absolute value signs are dropped in the formula for

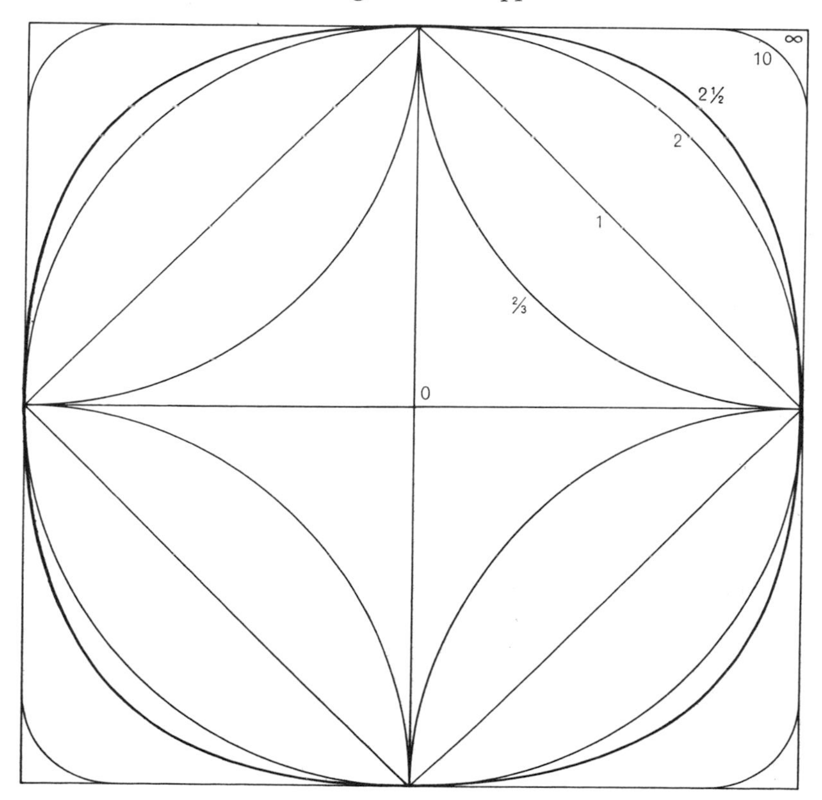

FIGURE 107
Supercircle and related curves

Lamé curves, you get a family of curves of which the famous Witch of Agnesi is a member. (The witch results when $n = 3$.) William Hogan wrote to say that parkway arches, designed by himself and other engineers, often are Lamé curves of exponent 2.2. In the thirties, he said, they were called "2.2 ellipses."

When a superellipse (a Lamé curve with exponent greater than 2) is applied to a physical object, its exponent and parameters a and b can, of course, be varied to suit circumstances and taste. For the Stockholm center, Piet Hein used the parameters $n = 2\frac{1}{2}$ and $a/b = 6/5$. A few years later Gerald Robinson, a Toronto architect, applied the superellipse to a parking garage in a shopping center in Peterborough, a Toronto suburb. The length and width were required to be in the ratio $a/b = 9/7$. Given this ratio, a survey indicated that an exponent slightly greater than 2.7 produced a superellipse that seemed the most pleasing to those polled. This suggested e as an exponent (since $e = 2.718 \ldots$). Robinson's use of e for the exponent has the consequence, writes Norman T. Gridgeman in his informative article on Lamé curves (see bibliography), that every point on the oval, except for the four points where it crosses an axis, is transcendental.

Readers suggested other parameters. J. D. Turner proposed mediating between the extremes of circle and square (or rectangle and ellipse) by picking the exponent that would give an area exactly halfway between the two extreme areas. D. C. Mandeville found that the exponent mediating the areas of a circle and square is so close to pi that he wondered if it actually *is* pi. Unfortunately it is not. Norton Black, using a computer, determined that the value is a trifle greater than 3.17. Turner also proposed mediating between ellipse and rectangle by choosing an exponent that sends the curve through the midpoint of a line joining the rectangle's corner to the corresponding point on the ellipse.

Turner and Black each suggested that the superellipse be combined with the aesthetically pleasing "golden rectangle" by making a/b the golden ratio. Turner's vote for the most pleasing superellipse went to the oval with parameters $a/b =$ golden ratio and $n = e$. Michel L. Balinski and Philetus H. Holt III, in a letter published in the *New York Times* in December, 1968 (I failed to record the day of the month), recommended a golden superellipse with $n = 2\frac{1}{2}$ as the best shape for the negotiating table in Paris. At that time the diplomats preparing to negotiate a Vietnam peace were quarreling over the shape of their table. If no table can be agreed upon, Balinski and Holt wrote, the diplomats should be put inside a hollow superegg and shaken until they are in "superelliptic agreement."

Sergel's Square, or Sergel's Torg as it is called in Sweden, is still under construction. The Superellipse Plaza, with its fountain pool on street level, has been completed. The Piet Hein Arcade below, with its shops and restaurant, is expected to be finished in 1979.

The superegg is a special case of the more general solid shape which one can call a superellipsoid. The superellipsoid's formula is

$$\left|\frac{x}{a}\right|^n + \left|\frac{y}{b}\right|^n + \left|\frac{z}{c}\right|^n = 1.$$

When $a = b = c$, the solid is a supersphere, its shape varying from sphere to cube as the exponent varies. When $a = b$, the solid is a superegg. Its formula is

$$\left|\frac{x}{a}\right|^n + \left|\frac{y}{a}\right|^n + \left|\frac{z}{b}\right|^n = 1;$$

which also can be written

$$\left|\frac{\sqrt{x^2 + y^2}}{a}\right|^n + \left|\frac{z}{b}\right|^n = 1.$$

When I wrote my column on the superellipse, I believed that any solid superegg based on an exponent greater than 2 and less than infinity would balance on its end provided its height did not exceed its width by too great a ratio. A solid superegg with an exponent of infinity would, of course, be a right circular cylinder that would, in principle, stand on its flat end regardless of how much higher it was than wide. But short of infinity it seemed intuitively clear that for each exponent there was a critical ratio beyond which the egg would be unstable. Indeed, I even published the following proof that this is the case:

If the center of gravity, CG, *of an egg is below the center of curvature,* CC, *of the egg's base at the central point of the base, the egg will balance. It balances because any tipping of the egg will raise the* CG. *If the* CG *is above the* CC, *the egg is unstable because the slightest tipping lowers the* CG. *To make this clear, consider first the sphere shown at the left in Figure 109. Inside the sphere the* CG *and* CC *are the same point: the center of the sphere. For any supersphere with an exponent greater than 2, as shown second from left in the illustration, the* CC *is above the* CG *because the base is less convex. The higher the exponent, the less convex the base and the higher the* CC.

Now suppose the supersphere is stretched uniformly upward along its vertical coordinates, transforming it into a superellipsoid of rotation, or what Piet Hein calls a superegg. As it stretches, the CC *falls and the* CG *rises. Clearly there must be a point* X *where the* CC *and the* CG *coincide. Before this crucial point is reached the superegg is stable, as shown third from left in Figure 109. Beyond that point the superegg is unstable (right).*

C. E. Gremer, a retired U.S. Navy commander, was the first of many readers to inform me that the proof is faulty. Contrary to intuition, at the base point of *all* supereggs, the center of curvature is infinitely high! If we increase the height of a superegg while its width remains constant, the curvature at the base point remains "flat." German mathematicians call it a *flachpunkt*. The superellipse has a similar *flachpunkt* at its ends. In other words, all supereggs, regardless of their height-width ratio, are theoretically stable! As a superegg becomes taller and thinner, there is of course a critical ratio at which the degree of

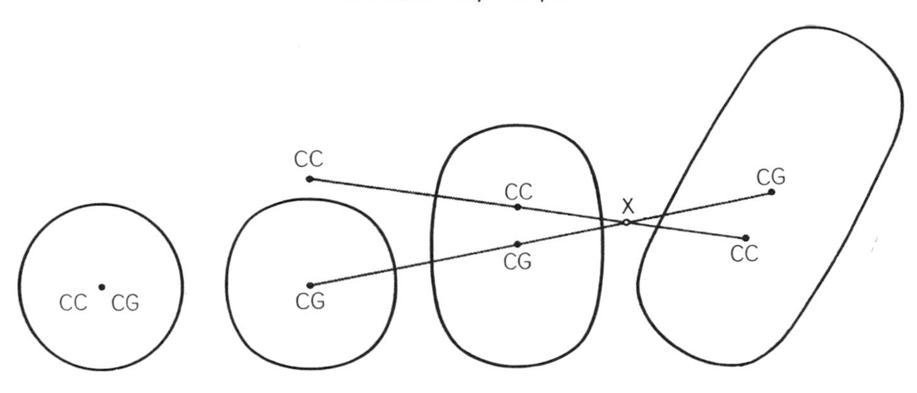

FIGURE 109
Diagrams for a false proof of superegg instability

tilt needed to topple it comes so close to zero that such factors as inhomogeneity of the material, surface irregularity, vibrations, air currents, and so on make it practically unstable. But in a mathematically ideal sense there *is* no critical height-width ratio. As Piet Hein has put it, in theory one can balance any number of supereggs, each an inch wide and as tall as the Empire State Building, on top of one another, end to end, and they will not fall. Calculating precise "topple angles" at which a given superegg will not regain balance is a tricky problem in calculus. Many readers tackled it and sent their results.

Speaking of egg balancing, the reader may not know that almost any chicken egg can be balanced on its broad end, on a smooth surface, if one is patient and steady-handed. Nothing is gained by shaking the egg first in an attempt to break the yolk. Even more puzzling as a parlor trick is the following method of balancing an egg on its *pointed* end. Secretly put a tiny amount of salt on the table, balance the egg on it, then gently blow away the excess grains before you call in viewers. The few remaining grains which hold the egg are invisible, especially on a white surface. For some curious reason, balancing chicken eggs legitimately on their broad ends became a craze in China in 1945—at least, so said *Life* in its picture story of April 9, 1945.

The world's largest superegg, made of steel and aluminum

and weighing almost a ton, was set up outside Kelvin Hall in Glasgow, October, 1971, to honor Piet Hein's appearance there as a speaker during an exhibition of modern homes. The superellipse has twice appeared on Danish postage stamps: In 1970 on a blue two-kroner honoring Bertel Thorvaldsen, and in 1972 on a Christmas seal bearing portraits of the queen and the prince consort.

Supereggs, in a variety of sizes and materials, are on sale throughout the world in stores that specialize in unusual gifts. Small, solid-steel supereggs are marketed as an "executive's toy." The best trick with one of them is to stand it on end, give it a gentle push, and try to make it turn one, two, or more somersaults before coming to rest again on one end. Hollow supereggs, filled with a special chemical, are sold as drink coolers. Larger supereggs are designed to hold cigarettes. More expensive supereggs, intended solely as art objects, are also being made. For information on how to obtain supereggs, as well as furniture, dishes, lamps, and other products based on a superelliptical design, write to the Piet Hein Information Center, Finsenvej 33,2000 Copenhagen F, Denmark.

CHAPTER 19

How to Trisect an Angle

TWO OF the first compass-and-straightedge constructions a child learns in plane geometry are the bisection of an angle and the division of a line segment into any desired number of equal parts. Both are so easy to do that many pupils find it hard to believe there is no way in which the two instruments can be used to trisect an angle. Indeed, it is usually the student who is most gifted mathematically who takes this as a challenge and immediately sets to work trying to prove the teacher wrong.

Something like this happened among mathematicians in the "childhood" of geometry. As far back as the fifth century B.C. geometers devoted a large share of their time to searching for a way of using straight lines and circles to obtain an intersection point that would trisect any given angle. They knew, of course, that certain angles could be trisected. The right angle, for instance, is ridiculously easy. One has only to draw the arc AB [*see Figure 110*] and then, without altering the compass opening, place the compass point at B and draw an arc that intersects the other arc at C. The line from O through C trisects the right angle. (Readers are invited to brush up on their plane geometry by proving the trisections mentioned in this article. All have simple proofs.) The 60-degree angle in turn trisects the straight angle of 180 degrees, and by bisecting the 30-degree angle one obtains the angle that trisects the 45-degree angle. An infinity of special angles obviously can be trisected under the classical restraints, but what the Greek geometers wanted was a general method applicable to any given angle. Together with doubling the cube and squaring the circle, finding it became one of the three great construction problems of ancient geometry.

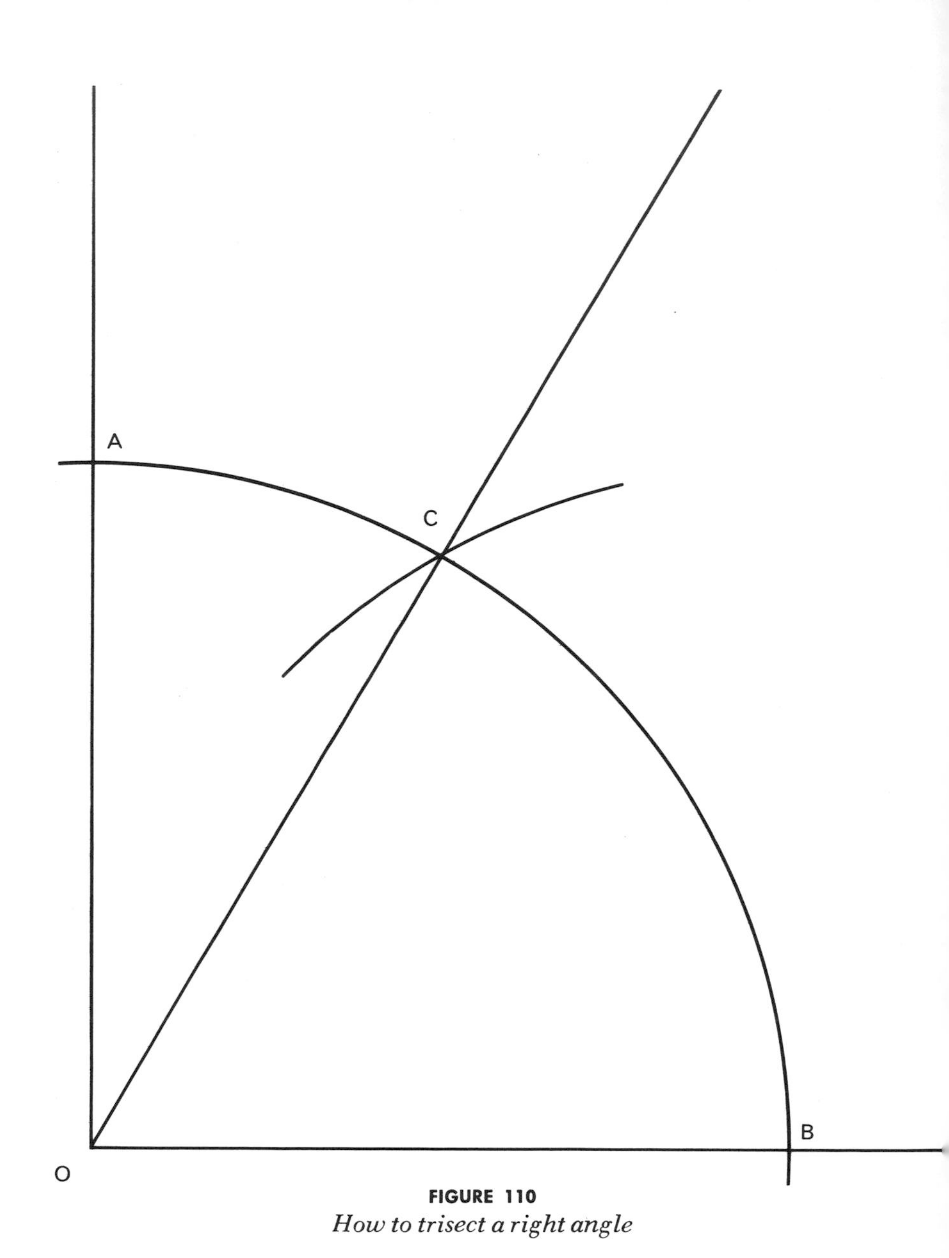

FIGURE 110

How to trisect a right angle

It was not until 1837 that a French mathematical journal published the first completely rigorous proof, by P. L. Wantzel, of the impossibility of trisection. His proof is much too technical to explain here, but the following remarks suggest its main lines. (The fullest and best nontechnical exposition of such a proof can be found in *What Is Mathematics?*, by Richard Courant and Herbert Robbins, pages 127–138.) Consider a 60-degree angle with its vertex at the origin of a Cartesian coordinate plane [*see Figure 111*]. Draw a circle with its center at *O* and assume that the circle's radius is 1. The trisecting line of the 60-degree angle will intersect this circle at *A*. Is it possible, using only compass and straightedge, to locate point *A*? If not, at least one angle cannot be trisected and therefore there will be no general method.

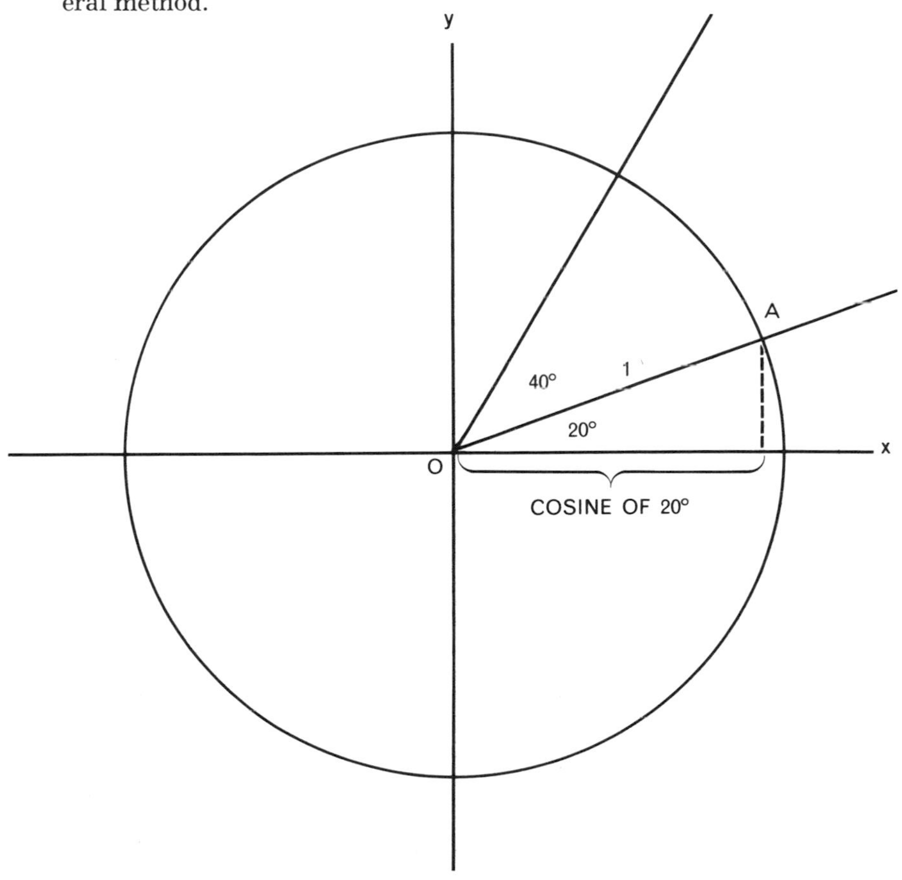

FIGURE 111
Point A is not "constructible" with compass and straightedge

Because straight lines on the Cartesian plane are graphs of linear equations and circles are graphs of quadratic equations, it can be shown that there are five, and only five, operations that can be performed on given line segments by using only a compass and straightedge. The segments can be added, subtracted, multiplied, and divided, and their square roots can be extracted. Given any line segment n, one can use compass and straightedge to find the square root of n. The same operation can be repeated on the square root of n to obtain *its* square root, which is the same as the fourth root of n. Thus by repeating the operation of square root extraction a finite number of times one can find any root in the doubling series 2, 4, 8, 16. . . . It is not possible, with compass and straightedge, to find the cubic root of any line segment, because 3 is not a power of 2. All of this, together with arguments in analytic geometry and the algebra of what are called "number fields," establishes that the only "constructible" points on the plane are those with x and y coordinates that are the real roots of a certain type of equation. It must be an algebraic equation that is irreducible (cannot be factored into expressions with lower exponents), has rational coefficients, and is of a degree that is a power of 2, that is, with its highest exponent in the 2, 4, 8 . . . doubling series.

Consider now the x coordinate of point A in Figure 111, the point that trisects the 60-degree angle. It measures the base of a right triangle whose hypotenuse is 1, and so it is equal to the cosine of 20 degrees. A bit of juggling with some simple trigonometric formulas shows that this cosine is the irrational root of an irreducible cubic equation: $8x^3 - 6x = 1$. The equation is of degree 3, therefore point A is not constructible. Since there is no way to find point A with compass and straightedge, the 60-degree angle cannot be trisected under the classic restrictions. Similar arguments prove there are no general methods by which compass and straightedge can divide any given angle into fifths, sixths, sevenths, ninths, tenths, or any other number of equal parts not in the 2, 4, 8, 16 . . . series. Among the infinity of angles that *can* be trisected are those equal in degrees to $360/n$

where n is an integer not evenly divisible by 3. Among the infinity that cannot be trisected are those equal to $360/n$ where n is an integer divisible by 3. An angle of 9 degrees can be trisected. Its trisector, the 3-degree angle, cannot—which is the same as saying that there is no way to construct a unit angle with compass and straightedge. Nor can a 2-degree angle be constructed.

There are, of course, many ways to trisect the angle approximately. One of the simplest (given by Hugo Steinhaus in his *Mathematical Snapshots*) is applied to a 60-degree angle in Figure 112. First the angle is bisected, then a chord of a half-angle is trisected. This provides a point for trisecting the original angle with an error less than the inevitable inaccuracy that occurs in the drawing. Dozens of still better approximations have been published, but most of them require considerably more work.

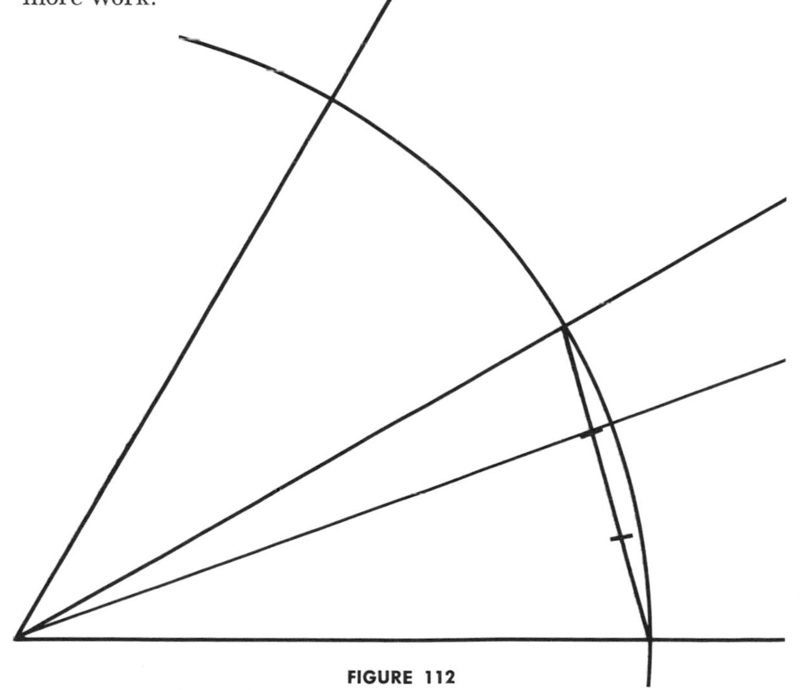

FIGURE 112
A simple way to (almost) trisect any angle

Absolutely precise trisections are achieved only by breaking one of the traditional restraints. Many noncircular curves such as the hyperbola and parabola will produce perfect trisections. Other methods assume an infinite number of construction steps with the trisecting line as a limit. The simplest way, however, to evade the restraints is to mark two points on the straightedge. This can even be done without actually putting marks on it: by using the ends of the straightedge to mark a line segment, by using its width or merely by pressing the legs of the compass firmly against the edge. One of the best trisections obtainable by this kind of cheating is found in the writings of Archimedes. The angle to be trisected is *AED* in Figure 113. Draw a semicircle as shown, then extend *DE* to the right. With the compass still open to the semicircle's radius, *DE*, hold the legs against the straightedge and place the straightedge so that it passes through point *A*. Adjust the straightedge until the points marked on it by the legs of the compass intersect the semicircle and its extended base at points *B* and *C*. In other words, make *BC* equal to the radius. The arc *BF* is now exactly one-third of the arc *AD*.

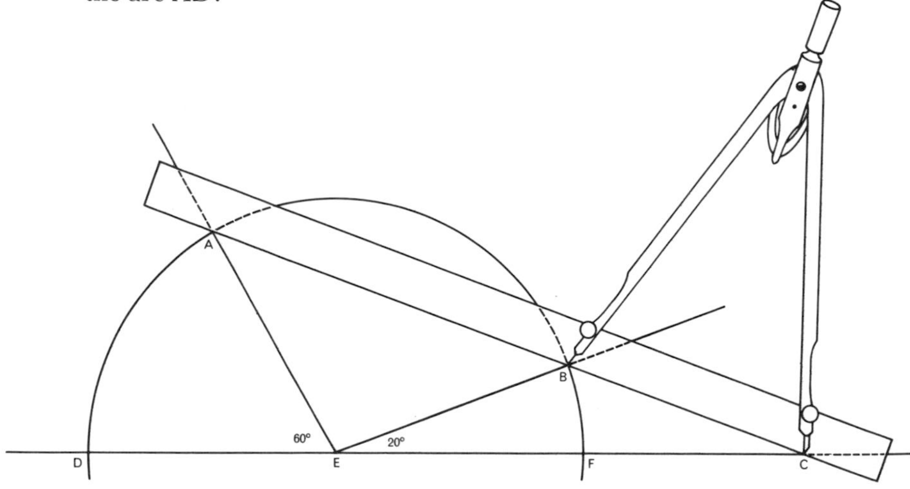

FIGURE 113
Archimedes' method of trisection by cheating

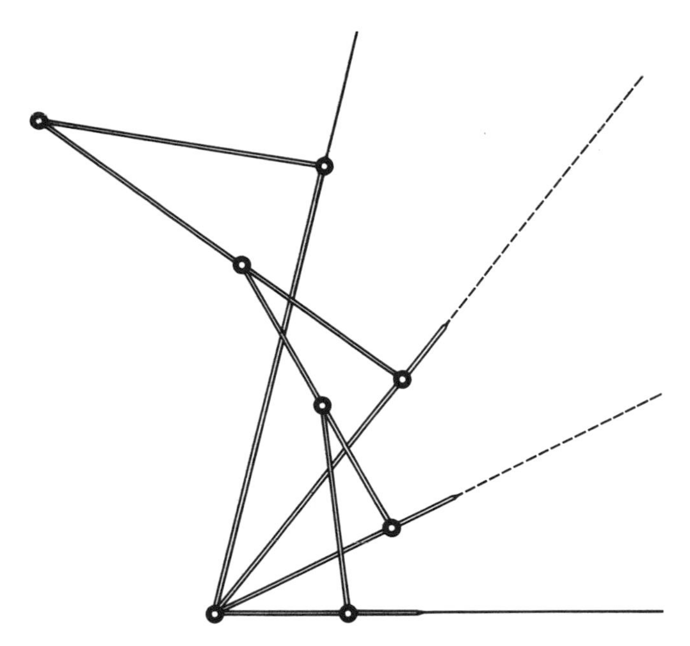

FIGURE 114
Kempe's linkage for trisecting any angle

A great variety of curious mechanical devices have been invented for angle-trisecting. (Twenty are pictured in the standard Italian work on recreational mathematics, *Matematica Dilettevole e Curiosa*, by Italo Ghersi, current edition, pages 476–489). As Leo Moser, a University of Alberta mathematician, once pointed out in an article, the ordinary watch is such an instrument. If the minute hand is moved over an arc equal to four times the angle to be trisected, the hour hand moves through an arc that is one-third the given angle. A whimsical linkage designed by Alfred Bray Kempe, a London lawyer, is based on theorems involving crossed parallelograms—parallelograms "folded" so that two opposite sides cross [*see Figure 114*]. The three crossed parallelograms in the linkage are similar. A long side of the smallest is a short side of the middle one, a long side of which is in turn a short side of the largest one. The device trisects automatically, as shown. The principle can be extended, by adding more crossed parallelograms, to make an instrument that will divide angles into any desired number of equal parts.

An easy-to-make cardboard trisecting device called the "tomahawk" has no moving parts, requires no preliminary construction lines, and is unconditionally guaranteed to trisect instantly and accurately [*see Figure 115*]. Its top edge *AD* is cut into thirds by points *B* and *C*. The curved edge is the arc of a semicircle with a radius *AB*. The tomahawk is placed with corner *D* on one arm of the angle, the semicircle tangent to the other arm, and the right edge of its handle crossing the angle's vertex. Points *B* and *C* trisect the angle. If an angle is too acute for the tomahawk to fit, you can always double it one or more times until it is large enough, trisect the larger angle, then halve the result as many times as you doubled it.

Although the proof of trisection impossibility by compass and straightedge is completely convincing to anyone who understands it, there are still amateur mathematicians all over the world who delude themselves into believing they have discovered a method that meets the classic requirements. The typical angle-trisector is someone who knows just enough plane geometry to work out a procedure but not enough to follow the impossibility proof or to detect the flaw in his own method. His trisection is often so complicated, and his proof has so many steps, that it is not easy for even the expert geometer to find the error that is certain to be there. Professional mathematicians are always being favored with such proofs. Since it is both time-consuming and unrewarding to search for errors, they usually mail the material back quickly without trying to analyze it. This invariably confirms the trisector's suspicion that the professionals are engaged in an organized conspiracy to prevent his great discovery from becoming known. After his method has been rejected by all the mathematical journals to which he sends it, he often explains it in a book or pamphlet printed at his own expense. Sometimes he describes his method in an advertisement in a local newspaper, adding that his manuscript has been properly notarized.

The last amateur mathematician to be given widespread publicity in the United States for angle-trisecting was the Very

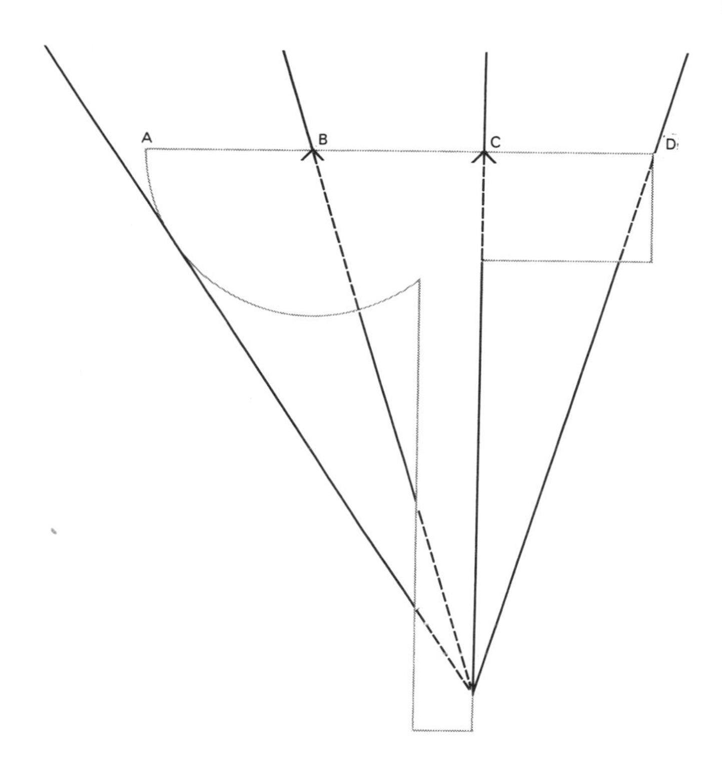

FIGURE 115
The "tomahawk" trisector

Reverend Jeremiah Joseph Callahan. In 1931, when he was president of Duquesne University in Pittsburgh, he announced that he had solved the trisection problem. The United Press sent out a long wire story, written by Father Callahan himself. *Time* ran his picture with a favorable account of his revolutionary discovery. (That same year Father Callahan published a 310-page book called *Euclid or Einstein*, in which he demolished relativity theory by proving Euclid's famous parallel postulate, thereby showing the absurdity of the non-Euclidean geometry on which general relativity is based.) Reporters and laymen expressed shock when the establishment mathematicians, *without even waiting to see Father Callahan's construction*, declared unequivocally that it could not be correct. Finally, at the end of the year, Duquesne published Father Callahan's booklet *The Trisection of the Angle.* "The mathematicians were right," says mathematician Irving Adler, who tells the story in his entertaining book *Monkey Business.* "Callahan had not trisected the angle." He had in effect merely taken an angle, tripled it, then found the original angle again.

On June 3, 1960, the Honorable Daniel K. Inouye, then a representative from Hawaii, later a senator and member of the Watergate investigation committee, read into the *Congressional Record* (Appendix, pages A4733–A4734) of the 86th Congress a long tribute to Maurice Kidjel, a Honolulu portrait artist who has not only trisected the angle but also squared the circle and duplicated the cube. Kidjel and Kenneth W. K. Young have written a book about it called *The Two Hours that Shook the Mathematical World*, and a booklet, *Challenging and Solving the Three Impossibles.* Through a company called The Kidjel Ratio they sell this literature along with the Kidjel ratio calipers with which one can apply the system. In 1959 the two men lectured on their work in a number of U.S. cities, and a San Francisco television station, KPIX, produced a documentary about them called *The Riddle of the Ages.* According to Inouye, "The Kidjel solutions are also now being taught in hundreds of schools and colleges throughout Hawaii, the United States, and Canada." One hopes his statement was exaggerated.

A correspondent in California sent me a clipping from the *Los Angeles Times* for Sunday, March 6, 1966 (Section A, page 16). A man in Hollywood had taken a two-column advertisement to display, in 14 steps, his method of trisecting.

What can the mathematician say today to an angle-trisector? He can remind him that in mathematics it is possible to define tasks that are impossible in a final, absolute sense: impossible at all times, in all conceivable (logically consistent) worlds. It is as impossible to trisect the angle as it is to move a queen in chess the way one moves a knight. In both cases the ultimate reason for the impossibility is the same: the operation violates the rules of a mathematical game. The mathematician can urge the trisector to get a copy of *What Is Mathematics?* and study the section mentioned above, then go back to his proof and try a bit harder to find where he went astray. But angle-trisectors are a persistent breed and not likely to take anyone's advice. Augustus De Morgan, in his *Budget of Paradoxes,* quotes a typical phrase from a 19th-century pamphlet on angle-trisection: "The consequence of years of intense thought." De Morgan's comment is terse: "Very likely, and very sad."

Bibliography

2. PENNY PUZZLES

Coin puzzles in general

Play Mathematics. Harry Langman. Hafner, 1962.
Fun for the Money. Maxey Brooke. Scribner's, 1963.

Isometric solitaire

"Triangular Puzzle Peg." Irvin Roy Hentzel, *Journal of Recreational Mathematics*, Vol. 6, Fall, 1973, pages 280–283.

Tree-plant problems

Mathematical Recreations and Essays. W. W. Rouse Ball and H. S. M. Coxeter. University of Toronto Press, twelfth edition, 1974, pages 104–105.
"Sylvester's Problem on Collinear Points." D. W. Crowe, *Mathematics Magazine*, Vol. 14, January, 1968, pages 30–34.

3. ALEPH-NULL AND ALEPH-ONE

Farey numbers

"Farey Sequences." George S. Cunningham, in *Enrichment Mathematics for High School*. National Council of Teachers of Mathematics, 1963, Chapter 1.
Recreations in the Theory of Numbers. Albert H. Beiler. Dover, 1964, Chapter 16.
Ingenuity in Mathematics. Ross Honsberger. Random House New Mathematical Library, 1970, Chapter 5.

Transfinite numbers

The Continuum. Edward V. Huntington. Dover, 1955.
Introduction to the Theory of Sets. Joseph Breuer. Prentice-Hall, 1958.
Sets and Transfinite Numbers. Martin M. Zuckerman. Macmillan, 1974.

Cohen's proof

Set Theory and the Continuum Hypothesis. Paul J. Cohen. W. A. Benjamin, 1966.

"Non-Cantorian Set Theory." Paul J. Cohen and Reuben Hersh, *Scientific American*, December, 1967, pages 104–116.

"The Continuum Problem." Raymond M. Smullyan, in *The Encyclopedia of Philosophy*, Macmillan, 1967.

"The Continuum Hypothesis." Raymond M. Smullyan, in *The Mathematical Sciences*, edited by the Committee on the Support of Research in the Mathematical Sciences (COSRIMS). M.I.T. Press, 1969, pages 252–271.

Alephs and cosmology

"Combinatorial Analysis in Infinite Sets and Some Physical Theories." S. Ulam, *SIAM Review*, Vol. 6, October, 1964, pages 343–355.

"The Problem of Infinite Matter in Steady-State Cosmology." Richard Schlegel, *Philosophy of Science*, Vol. 32, January, 1965, pages 21–31.

Completeness in Science. Richard Schlegel. Appleton-Century-Crofts, 1967, pages 138–149.

4. HYPERCUBES

A New Era of Thought. C. Howard Hinton. Swan Sonnenschein, 1888.

The Fourth Dimension. C. Howard Hinton. Allen & Unwin, 1904.

The Fourth Dimension. E. H. Neville. Cambridge University Press, 1921.

Geometry of Four Dimensions. Henry Parker Manning. Dover, 1956.

Christian Faith and Natural Science. Karl Heim. Harper Torchbook, 1957.

"The Ifth of Oofth." Walter S. Tevis, Jr., *Galaxy Science Fiction*, April, 1957, pages 59–69. A wild and funny tale about a fifth dimensional cube that distorts space and time.

An Introduction to the Geometry of N Dimensions. D. M. Y. Sommerville. Dover, 1958.

The Fourth Dimension Simply Explained. Henry Parker Manning. Dover, 1960.

Regular Polytopes. H. S. M. Coxeter. Dover, third edition, 1973.

5. MAGIC STARS AND POLYHEDRONS

Magic Squares and Cubes. W. S. Andrews. Dover Publications, 1960, Chapter XIII.

Play Mathematics. Harry Langman. Hafner Publishing Co., 1962, Chapter VI.

536 Puzzles and Curious Problems. H. E. Dudeney. Scribner's, 1967, pages 145–147.

6. CALCULATING PRODIGIES

Additional references may be found in William L. Schaaf, *A Bibliography of Recreational Mathematics*, Vol. 1, National Council of Teachers of

Mathematics, fourth edition, 1970, pages 39–42, and in the extensive European bibliography in Regnault's book, listed below.

"On Mental Calculation." George P. Bidder, *Proceedings of the Institute of Civil Engineers,* Vol. 15, 1856, pages 251–280.

"Arithmetical Prodigies." E. W. Scripture, *American Journal of Psychology,* Vol. 4, April, 1891, pages 1–59.

Psychologie des grandes calculateurs et joueurs d'échecs. A. Binet. Paris: Hachette, 1894.

"Calculating Boys." Anonymous, *Strand Magazine,* Vol. 10, 1895, pages 277–280.

"Mathematical Prodigies." Frank D. Mitchell, *American Journal of Psychology,* Vol. 18, January, 1907, pages 61–143.

"Lightning Calculators." H. Addington Bruce, *McClure's Magazine,* Vol. 39, 1912, pages 586–596.

"Arithmetical Prodigies." R. C. Archibald, *American Mathematical Monthly,* Vol. 25, 1918, pages 91–94.

"Visual Imagery of a Lightning Calculator." W. A. Bausfield and H. Barry, *American Journal of Psychology,* Vol. 45, 1933, pages 353–358.

"Examination of the Computing Ability of Mr. Salo Finkelstein." J. D. Weinland and W. S. Schlauch, *Journal of Experimental Psychology,* Vol. 21, October, 1937, pages 382–402.

"Memory of Salo Finkelstein." J. D. Weinland, *Journal of General Psychology,* Vol. 39, October, 1948, pages 243–257.

"A Medical-psychological Account Followed by a Demonstration of a Case of Super-normal Aptitude." B. Stovkis, *Proceedings of the International Congress on Orthopedagogics,* Amsterdam, July, 1949. On the Klein brothers.

Mental Prodigies. Fred Barlow. Philosophical Library, 1952.

Les Calculateurs Prodiges. Jules Regnault. Paris: Payot, 1943, revised edition, 1952.

"The Art of Mental Calculations; With Demonstrations." A. C. Aitken, *Transactions of the Society of Engineers,* Vol. 44, December, 1954, pages 295–309.

"Calculating Prodigies." W. W. Rouse Ball and H. S. M. Coxeter, in *Mathematical Recreations and Essays.* University of Toronto Press, twelfth edition, 1974.

"An Exceptional Talent for Calculative Thinking." Ian M. L. Hunter, *British Journal of Psychology,* Vol. 53, Part 3, August, 1962, pages 243–258. On Aitken.

The Magic of Numbers. Robert Tocquet. Premier Books, 1962.

"Strategies for Skill." Ian M. L. Hunter, in *Penguin Science Survey 1965B.* Penguin Books, 1965. On Aitken.

"Alexander Craig Aitken: New Zealand's Greatest Mathematician." H. P. Kidson. *The New Zealand Mathematics Magazine,* Vol. 10, November, 1973, pages 129–133.

7. TRICKS OF LIGHTNING CALCULATORS

"Arthur Griffith, Arithmetical Prodigy." William L. Bryan, Ernest H. Lindley, and Noble Herter, in *On the Psychology of Learning a Life Occupation.* Indiana University, 1935, pages 11–65.

Les Calculateurs Prodiges. Jules Regnault. Paris: Payot, 1943, revised edition, 1952.

An Adventure in Figuring: The Barrett Blitz Method of Extracting Cube Root. Urbane L. Barrett. Los Angeles, privately published, 1943. An excellent technique for rapid mental calculation of cube roots greater than 100.

Math Miracles. Wallace Lee. Privately published, 1950, revised edition, 1960.

Mental Prodigies. Fred Barlow. Philosophical Library, 1952.

Mathematics, Magic and Mystery. Martin Gardner. Dover, 1956, Chapter 9.

The Magic of Numbers. Robert Tocquet. Premier Books, 1962.

The calendar trick

There are hundreds of references to books and magazine articles giving methods and formulas for determining the day of the week for a given date. (See William Schaff, *Bibliography of Recreational Mathematics,* Vol. 1, fourth edition, 1970, pages 30–31; Vol. 2, fourth edition, 1970, pages 4–5.) I list below only references of special interest to the lightning calculator.

Roth Memory Course. David M. Roth. Writers Publishing Co., 1918, revised, 1965, pages 252–263.

Calendar Memorizing. Bernard Zufall. Privately printed pamphlet, 1940.

Math Miracles. Wallace Lee. Privately published, 1950, revised, 1960, Chapter 19.

"A Mental Calendar." Rev. Brother Leo, *The Mathematics Teacher,* Vol. 50, October, 1957, pages 438–439.

"Tomorrow is the Day After Doomsday." John H. Conway, *Eureka,* No. 36, October, 1973, pages 28–31. A novel system derived from Lewis Carroll's rule.

8. THE ART OF M. C. ESCHER

Symmetry Aspects of M. C. Escher's Periodic Drawings. Caroline H. MacGillavry. A. Oosthoek's Uitgeversmaatschappij NV, 1965.

The Graphic Work of M. C. Escher. Duell, Sloan and Pearce, revised edition, 1967.

The World of M. C. Escher. Edited by J. L. Locher. Harry N. Abrams, 1971.

"Escher: The Journey to Infinity." Ken Wilkie, *Holland Herald,* Vol. 9, No. 1, 1974, pages 20–43.

"Master of Tesselations: M. C. Escher, 1898–1972." Ernest R. Ranucci, *Mathematics Teacher,* Vol. 67, April, 1974, pages 299–306.

"How to Draw Tesselations of the Escher Type." Joseph L. Teeters, *Mathematics Teacher*, Vol. 67, April, 1974, pages 307–310.

"Sources of Ambiguity in the Prints of Maurits C. Escher." Marianne L. Teuber, *Scientific American*, Vol. 231, July, 1974, pages 90–104. See also the correspondence on this article in Vol. 232, January, 1975, pages 8–9.

10. CARD SHUFFLES

By mathematicians

Elementary Number Theory. J. V. Uspensky and M. A. Heaslat. McGraw-Hill, 1939.

"Congruences and Card Shuffling." Paul B. Johnson. *American Mathematical Monthly*, Vol. 63, December, 1956, pages 718–719.

An Introduction to Probability Theory and Its Applications, Vol. 1. William Feller. John Wiley & Sons, 1950, pages 367–372.

"Permutations by Cutting and Shuffling." S. W. Golomb, *SIAM Review*, Vol. 3, October, 1961, pages 293–297.

The Theory of Gambling and Statistical Logic. Richard A. Epstein. Academic Press, 1967, pages 181–193.

"Make Up Your Own Card Tricks." Irving Adler, *Journal of Recreational Mathematics*, Vol. 6, Spring, 1973, pages 87–91.

Matters Mathematical. I. N. Herstein and I. Kaplansky. Harper and Row, 1974, pages 118–121.

By magicians

"Trailing the Dovetail Shuffle to Its Lair." Charles T. Jordan, *The Bat*, November, December, 1948; January, February, March, 1949.

Expert Card Technique. Jean Hugard and Frederick Braue. Faber and Faber, 1954, Chapter 16.

"The Mathematics of the Weave Shuffle." Alex Elmsley, *The Pentagram*, Vol. 11, June, 1957, pages 70–71; July, 1957, pages 78–79; August, 1957, page 85; Vol. 12, May, 1958, page 62.

The Faro Shuffle. Edward Marlo. Chicago: Privately published, 1958.

Faro Notes. Edward Marlo. Chicago: privately published, 1958.

Faro Controlled Miracles. Edward Marlo. Chicago: privately published, 1964.

Faro Possibilities. Karl Fulves. Teaneck, New Jersey: privately published, 1966.

Faro Fantasy. Paul Swinford. Connersville, Ind.: privately published, 1968.

More Faro Fantasy. Paul Swinford. Connersville, Ind.: privately published, 1971.

Faro and Riffle Technique. Karl Fulves. Teaneck, New Jersey: privately published, 1974.

"A Solution to Elmsley's Problem." Murray Bonfeld, *Genii*, May, 1973, pages 195–196.

"A Name Revelation with Faro Shuffles." Michael S. Ewer, *Genii*, November, 1973, pages 465–468.

11. MRS. PERKINS' QUILT AND OTHER SQUARE-PACKING PROBLEMS

Cyclopedia of Puzzles. Sam Loyd. Lamb Publishing Co., 1914, pages 39, 65.

Amusements in Mathematics. H. E. Dudeney. Thomas Nelson and Sons, 1917, Problem 173.

Puzzles and Curious Problems. H. E. Dudeney. Thomas Nelson and Sons, 1931, Problem 177.

"Mrs. Perkins's Quilt." J. H. Conway, *Proceedings of the Cambridge Philosophical Society*, Vol. 60, July, 1964, pages 363–368.

"Mrs. Perkins's Quilt." G. B. Trustrum, *Proceedings of the Cambridge Philosophical Society*, Vol. 61, January, 1965, pages 7–11.

"Some Packing and Covering Theorems." J. W. Moon and L. Moser, *Colloquium Mathematicum*, Vol. 17, 1967, pages 103–110.

"On Packing of Squares and Cubes." A. Meir and L. Moser, *Journal of Combinatorial Theory*, Vol. 5, September, 1968, pages 126–134.

12. THE NUMEROLOGY OF DR. FLIESS

Studies in the Psychology of Sex. Havelock Ellis. Random House, 1936. (See index references to Fliess.)

The Life and Work of Sigmund Freud, Vol. 1. Ernest Jones. Basic Books, 1953, Chapter 13.

The Origins of Psycho-Analysis: Letters to Wilhelm Fliess. Edited by Marie Bonaparte, Anna Freud, and Ernst Kris. Basic Books, 1954.

Biorhythm. Hans J. Wernli. Crown, 1961.

Is This Your Day? George Thommen. Crown, 1964.

13. RANDOM NUMBERS

"Order and Surprise." Martin Gardner, *Philosophy of Science*, Vol. 17, January, 1950, pages 109–117.

"A Statistical Study of Randomness Among the First 10,000 digits of Pi." *Mathematics of Computation*, Vol. 16, April, 1962, pages 188–197.

"Random Number Generators." T. E. Hull and A. R. Dobell, *SIAM Review*, Vol. 4, July, 1962, pages 230–254.

"Randomness." W. Allen Wallis and Henry V. Roberts in *The Nature of Statistics*. Collier Books, 1962, Chapter 6.

Random Number Generators. Birger Jansson. Stockholm: Almqvist and Wiksell, 1966.

"Randomness and the Twentieth Century." Alfred M. Bork, *Antioch Review*, Vol. 27, Spring, 1967, pages 40–61.

"Random Numbers." *Seminumerical Algorithms*, Chapter 3. Donald E. Knuth, Addison-Wesley, 1969.

15. PASCAL'S TRIANGLE

Articles on Pascal's triangle, giving generalizations, variations, properties, and applications, are so numerous that I will list here only a few that have appeared since 1970. For a selected list of earlier references, see William L. Schaaf, *A Bibliography of Recreational Mathematics,* Vol. 2, National Council of Teachers of Mathematics, 1970, pages 21–22.

"Spaces, Functions, Polygons, and Pascal's Triangle." Lars C. Jansson, *Mathematics Teacher*, Vol. 66, January, 1973, pages 71–77.

"Pascal's Triangle Revisited." James K. Bidwell, *Mathematics Teacher*, Vol. 66, May, 1973, pages 448–452.

"Perfect Square Patterns in the Pascal Triangle." Zalman Usiskin, *Mathematics Magazine*, Vol. 46, September, 1973, pages 203–208.

16. JAM, HOT, AND OTHER GAMES

"The Game is Hot." Leo Moser, *Recreational Mathematics Magazine*, Vol. 1, June, 1961, pages 23–24.

Differential Games. Rufus Isaacs. Wiley, 1965.

Nim-like games

"The *G*-Values of Various Games." Richard K. Guy and Cedric A. B. Smith, *Proceedings of the Cambridge Philosophical Society*, Vol. 52, Part 2, July, 1956, pages 514–526.

"Disjunctive Games with the Last Player Losing." P. M. Grundy and Cedric A. B. Smith, *Proceedings of the Cambridge Philosophical Society*, Vol. 52, Part 2, July, 1956, pages 527–533.

Mathematical Recreations and Essays. W. W. Rouse Ball and H. S. M. Coxeter. University of Toronto Press, twelfth edition, 1974, pages 36–40.

The Theory of Gambling and Statistical Logic. Richard A. Epstein. Academic Press, 1967, Chapter 10.

"Compound Games with Counters." Cedric A. B. Smith, *Journal of Recreational Mathematics*, Vol. 1, April, 1968, pages 66–77.

17. COOKS AND QUIBBLE-COOKS

Amusements in Mathematics. H. E. Dudeney. Dover, 1958.

536 Puzzles and Curious Problems. H. E. Dudeney. Scribner's, 1967. This reprints two of Dudeney's books: *Modern Puzzles* and *Puzzles and Curious Problems.*

Mathematical Puzzles of Sam Loyd, Vols. 1 and 2. Edited by Martin Gardner. Dover, 1959, 1960.

18. PIET HEIN'S SUPERELLIPSE

The superellipse

"Note on Squares and Cubes." J. Allard, *Mathematics Magazine,* Vol. 37, September, 1964, pages 210–214.

"Lamé Ovals." Norman T. Gridgeman, *Mathematical Gazette,* Vol. 54, February, 1970, pages 31–37.

Piet Hein

"Piet Hein Bestrides Art and Science." Jim Hicks, *Life,* October 14, 1966, pages 55–66.

"King of Supershape." Anne Chamberlin, *Esquire,* January, 1967, page 112*ff*.

19. HOW TO TRISECT AN ANGLE

The Trisection Problem. Robert Carl Yates. Franklin Press, 1942; National Council of Teachers of Mathematics, 1971.

Number Theory and Its History. Oystein Ore. McGraw-Hill, 1948, Chapter 15.

Famous Problems of Elementary Geometry. Felix Klein. Dover, 1956, Chapter 2.

A Long Way from Euclid. Constance Reid. T. Y. Crowell, 1963, Chapter 9.

Famous Problems of Mathematics. Heinrich Tietze. Graylock Press, 1965, Chapter 3.